Photo-switched Biodegradation of Bioplastics in Marine Environments

Tatsuo Kaneko
Editor

Photo-switched Biodegradation of Bioplastics in Marine Environments

Springer

Editor
Tatsuo Kaneko
School of Chemical and Material Engineering
Jiangnan University
Wuxi, China

Graduate School of Advanced Science and Technology
Japan Advanced Institute of Science and Technology
Nomi, Ishikawa, Japan

ISBN 978-981-99-4353-1 ISBN 978-981-99-4354-8 (eBook)
https://doi.org/10.1007/978-981-99-4354-8

This book is based on results obtained from a Moonshot R&D project, "Development of photoswitching ocean-degradable plastics with edibility", JPNP18016, commissioned by the New Energy and Industrial Technology Development Organization (NEDO).

This Springer imprint is published by the registered company Springer Nature Singapore Pte Ltd.
The registered company address is: 152 Beach Road, #21-01/04 Gateway East, Singapore 189721, Singapore

Preface

We are using various soft materials such as paper, cloth, wood, leather, film, rubber, and sponges, whose main constituents are polymers. Polymers are composed of macromolecules which are long, thread-like organic molecules. Plastics are also composed of macromolecules. Other soft materials include food and ornamental plants, and then our bodies are also made up of macromolecules such as proteins and DNA which should be biodegradable molecules. Plastics history starts from the modification of natural resins and materials. However, non-degradable plastics such as polyolefin became so convenient that they later began to replace naturally occurring resins and became indispensable to mankind society. Based on the replacement, various problems have arisen such as the marine plastic waste problem. Bio-based and biodegradable plastics introduced in this book have been developed to solve these problems. At present, such plastics do not have a variety for meeting the society demands, and we are still waiting for plastics whose degradability can be controlled. In other words, there is a need for high-performance plastics that do not degrade when used in the society, but transform into biodegradable plastics when disposed by the action of the natural environment as a degradation-switch. In particular, photonic switches are attracting attention as a solution to this problem. The following two basic systems are proposed for photoswitchable degradable plastics.

(1) ON-type photoswitch: This is a "photoswitch" in which sunlight reaches the inside of the plastic and starts biodegradation (ON) due to surface damage that occurs in the process from disposal to discharge into the ocean, although the plastic is stable as a material in the terrestrial life zone. Plastic products that dissipate in the terrestrial environment and can be discharged into the ocean, especially those with low specific gravity, those containing air, and fibrous materials with high surface tension (e.g., microfibers), are applicable.
(2) OFF-type light switch: This is a "photoswitch" in which biodegradation is inhibited (turned off) in the presence of fluorescent lights and sunlight exposure, and biodegradation begins in dark environments such as underwater, on the sea floor, and in composting. Since this photoswitch is applied to conventional marine degradable plastics, it can be applied to fishery materials, agricultural materials, and leisure goods. In particular, those with specific gravity greater than 1 or those integrated with heavy substances will be addressed.

The above two technologies are photoswitches of different dimensions and their combinations enable control of optical switches for various applications. In addition, we plan to propose an ON/OFF equipped optical switch that combines these two technologies as an advanced system.

This book describes the innovative materials in science and technology that have been applied for the development of photo-switchable biodegradable plastics in terms of raw bioresource development, polymerization technology, molding processing, photocatalyses and their composites with biodegradable polymers, marine-degradation evaluation in lab and actual area of ocean, enzymatic and microorganismal degradation, and toxicity evaluation methodology development. Actually it takes a long time to complete the works with such wonderful topics and still on the way. Then we put together the current technologies of individual elements to achieve this goal in this book.

Wuxi, China
Ishikawa, Japan

Tatsuo Kaneko
tkaneko@jiangnan.edu.cn

Contents

Introduction—Next Generation Biodegradable Plastics

Tatsuo Kaneko

1 General Introduction

If you look around at the materials around you and search for something relatively soft among them, you will find paper, cloth, wood, leather, film, rubber, and sponges (Fig. 1). The main constituents of all of these materials are macromolecules, which are long, thread-like organic molecules. Plastics are also composed of macromolecules, and are called soft materials in contrast to hard materials such as ceramics and metals. In fact, if you look at a section of plastic under a microscope with a very high magnification rate, such as an electron microscope, you can observe a thread-like substance. Other soft materials include food and ornamental plants. They are also composed of polymers. We ourselves are also made up of macromolecules such as proteins and DNA.

How did these macromolecules come into being naturally? There is still no clear answer to this question, and there are many theories about the process by which the first molecules were formed. Stanley Miller proved that amino acids were formed inanimate when lightning struck the mixture of water, methane, ammonia, and hydrogen that spontaneously existed in the atmosphere of the primitive earth, as shown in Fig. 2 [1], and later the formation of nucleic acids, the building blocks of DNA, was also confirmed. There is a theory that such a phenomenon is the origin of the birth of organic matter, but it is now considered to be a past theory because the above atmospheric composition is considered to be erroneous. There is also a theory that amino acids and other substances born in some environment in outer space were carried to the earth by meteorites [2]. Furthermore, it is considered difficult to obtain sufficient concentrations of amino acids according to these theories, and there is another theory that inorganic substances that have been gushing out from hydrothermal vents on the ocean floor for tens of thousands of years react in the micropores of minerals that act as catalysts and synthesize organic substances [3].

In any case, on the ancient earth, basic molecules such as amino acids, nucleic acids, sugars, and lipids were born, and through some interaction or the influence of the external environment, they became a droplet-like aggregate called a coacervate, which is believed to have been the source of life. In other words, molecules came into close contact with each other, and when the conditions were right,

Prof. T. Kaneko
School of Chemical and Material Engineering, Jiangnan University, Wuxi 214122, Jiangsu, China

Prof. T. Kaneko (✉)
Graduate School of Advanced Science and Technology, Japan Advanced Institute of Science and Technology, Nomi, Ishikawa 923-1211, Japan
e-mail: tkaneko@jiangnan.edu.cn

T. Kaneko (ed.), *Photo-switched Biodegradation of Bioplastics in Marine Environments*,
https://doi.org/10.1007/978-981-99-4354-8_1

Fig. 1 Photograph of the author's office. The main materials are described. Except for metal and glass (in red), everything else is made of polymers (in blue). There really are a lot of polymers around us

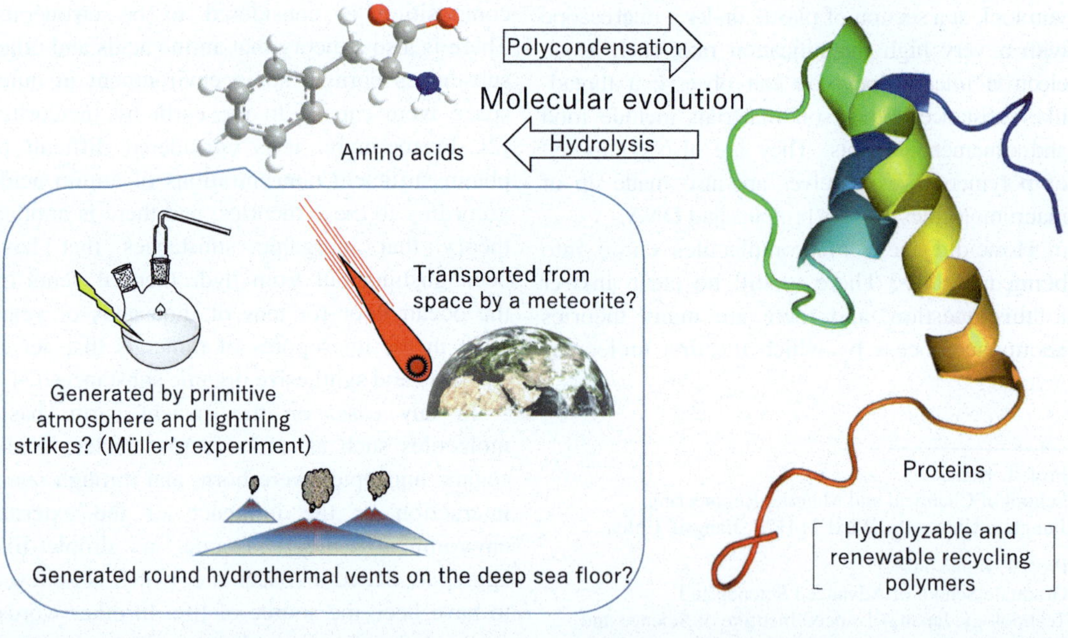

Fig. 2 Amino acids originated in ancient times and then developed naturally into proteins by molecular evolution. Proteins can be hydrolyzed to amino acids by natural processes, and the amino acids are regenerated from them. Therefore, protein is a resource-recycling macromolecule

chemical reactions took place, new molecules were born, and molecular evolution took place. Such chemical reactions have been confirmed to occur near hydrothermal vents, and of course they also occur inside living organisms. When one chemical reaction is taking place, another chemical reaction is taking place right next to it. As these chemical reactions occur in parallel, more stable molecules are produced by the chemical reactions that are more likely to occur. In other words, chemical reactions and molecules are also selected naturally depending on the environment. At this time, a molecule A reacts chemically with another molecule B to produce a completely different molecule C, which is then used in a chemical reaction with another molecule to produce a molecule D, which in turn produces a molecule E, etc. Chain reactions can occur. If an efficient chain reaction occurs, molecules join together in a thread-like structure to form a macromolecule (in living organisms, the chain reaction is controlled by enzymes). Although the detailed evolutionary process is unknown, the molecules formed by the chain reaction in such a long thread-like chain are polymers such as DNA (deoxyribonucleic acid), RNA (ribonucleic acid), proteins, and polysaccharides. On the contrary, the structure of natural macromolecules includes the process of molecular evolution to some extent. For example, when a protein is hydrolyzed, the amino acids that make up the protein can be obtained. This hydrolysis is a reverse reaction of the chemical reaction described above, and this reverse reaction is always carried out and recycled in living organisms. In other words, the created macromolecules are never left in the cell. The extracellular secreted molecules and dead cells are left in the environment, but they do not accumulate in the environment because various actions in other organisms and the environment trigger the decomposition reaction. Namely, only circulating molecules that can be degraded in the natural environment (not only biodegradation but also photooxidative degradation is a kind of environmental degradation) have been formed through molecular evolution. Molecules have been circulating in the biological world for hundreds of millions of years. These molecules and their constituent substances rarely kill other organisms unintentionally.

Plastics are also materials composed of polymers, but they are artificial polymers created by humans, and most of them are designed without considering molecular evolution. These plastics are scattered on the ocean surface for long periods of time, and minute fragments can be swallowed by marine organisms and cause death, while fishing nets can entangle marine organisms and cause them to suffer until they die. Next-generation plastics must be sustainable, and environmentally non-degradable materials that unintentionally destroy ecosystems will be eliminated in a future society. In other words, next-generation plastics should be developed based on molecular design in line with the direction of biological evolution, and should safely decompose in the natural world. In the case of genetically modified organisms, there are concerns about artificially controlling biological evolution, and ethical issues have been pointed out due to the fear of evolving in the wrong direction, but this level of consideration has not been taken into account when synthesizing various artificial molecules, so that molecular evolutions of some plastics have taken place in the wrong direction. Such plastic materials need to be replaced by environmentally degradable materials.

It is said that ancestor apes were born about several million years ago [4] and came to use tools, but at first they used materials found in nature as they were or processed them to make them easier to use. (Hard materials such as stone tools can also be said to be composed of polymers in the broad sense of the term, but they are excluded from this book because their concept is far removed from that of this book.) The prototype of plastic is a natural resin called shellac, which is extracted from the scale insect, and there is a record that it was used as a raw material for music records [5]. In other words, plastic originally started from the utilization of a natural substance itself, a resin that decomposes naturally. Later, celluloid (a composite of nitrated cellulose and camphor) was invented by Hyatt [6] and billiard balls were manufactured in earnest in 1869. This is said to be the first example of the industrialization of thermoplastic. Until then, billiard balls were made from ivory, but since only a maximum of eight billiard balls could be made from the ivory of one elephant, many elephants were overhunted. The first plastic, thus, had ecological significance. As its name suggests, celluloid is a plant-derived plastic (semi-synthetic plastic, not completely derived from plants) made from cellulose, and the cellulose part and camphor are naturally degradable. Although celluloid has been replaced by stable synthetic plastics due to its instability, the original plastic was naturally degradable (Fig. 3).

Here, in light of the molecular evolution mentioned earlier, biodegradable plastics that decompose naturally are considered to have inherited the molecular evolution inherited from ancient times, and are considered to be appropriate as next-generation plastics. The development of plastics using natural polymers such as cellulose is one direction, and many plastics derived from natural polymers have been developed so far, including cellophane. Cellulose is a polymer classified as a polysaccharide, which evolved from glucose and other sugars connected in long threads. Polysaccharides are the most abundant polymers on the earth, and their utilization is an extremely important molecular design guideline. It would be good if polysaccharides could be used as they are, like cellulose-rich cotton is used as dough, but most of them are highly water-soluble and difficult to handle. Even if starch is made into a film, it can only be used to make oblate, which has limited applications and is far from the image of plastic. Therefore, there has been a research trend to develop semi-synthetic plastics by chemical modification of polysaccharides (nitration for celluloid, acetation, ethylation, hydroxypropylation, carboxymethylation, etc.). For example, carboxymethylated cellulose is well known as CMC [7], a material used in laundry rinses. This trend is expected to continue into the next generation. Similarly, it is now possible to obtain DNA [8] and proteins [9] plastic films by chemical modification or special processing techniques. In addition, there is a polyester called polyhydroxyalkanoate (PHA) [10], which is accumulated by microorganisms as a storage carbon source, and its application as a biodegradable plastic is being studied.

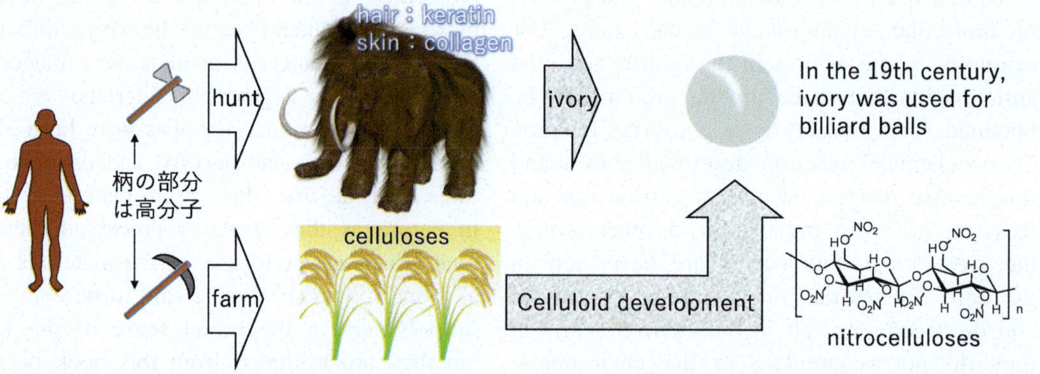

Since ancient times, human beings have made tools by combining macromolecular materials such as wood (cellulose, lignin, etc.) and stones, obtained food through hunting and farming, and developed and utilized macromolecular materials such as fur and rice straw.

In 1869, celluloid was made from cellulose nitrate and camphor, and used as a substitute for ivory. The first plastic was a bioplastic that saved the lives of many elephants.

Fig. 3 Organic materials used as human tools and the first plastics

These are considered to be polymers designed by utilizing the flow of molecular evolution. On the other hand, there is a polymer design that does not utilize molecular evolution but the polymer does not accumulate in the environment by providing biodegradability, thereby setting it in the same direction as the flow of molecular evolution. This is the trend of synthetic biodegradable plastics such as poly(lactic acid) [11]. Lactic acid is a natural molecule formed by molecular evolution, but poly(lactic acid), a type of polyester obtained by its dehydration reaction, does not exist in nature. Nevertheless, it is a biodegradable plastic that slowly decomposes into carbon dioxide and water under the action of nature. This is the trend in the development of next-generation plastics that should anticipate molecular evolution. Moreover, even if they are scattered as garbage, they will gradually disappear, so it will be possible to considerably suppress the unintentional killing of living organisms, such as ghost fishing. This trend includes design guidelines for the development of plastics that mimic the structure of natural polymers but with higher performance and functionality. Some of these technologies are based on wisdom that has a strong influence on the sustainability of the human race. One thing should be noted here. There is something called biomass plastic that is often confused with biodegradable plastic (Fig. 4). The Japan BioPlastics Association defines biomass plastic as "a polymer material containing substances derived from renewable organic resources as raw materials and obtained through chemical or biological synthesis. In particular, plastics with a biomass plastic content of 25% or more of the product weight are identified as biomass plastics" [12]. To calculate the biomass ratio, the content of carbon isotope C14 is quantified by accelerator mass spectrometry (AMS) or other methods [13]. The carbon isotope C14 is rarely contained in underground resources such as petroleum, but a certain amount is contained in biomass materials, making it possible to calculate. By recommending biomass plastics, we aim to have an effect on the reduction of carbon dioxide emissions. Recently, in response to the growing problem of marine plastic waste, Japan has started to charge for plastic shopping bags, and plastic shopping bags with a biomass plastic content of 25% or more are excluded from the charge. This is a typical example of the confusion between biodegradable plastics and biomass plastics.

The raw material of polyethylene that makes up plastic bags, is ethylene gas, which is also a biomolecule that promotes the ripening of apples and other fruits. It can also be synthesized by dehydration reaction of ethanol obtained by fermentation, which can be polymerized to obtain polyethylene that is certified as a biomass plastic. However, as long as it is polyethylene, it does not biodegrade. In the latter half of the twentieth century, it was argued that polyethylene would continue to degrade in the environment and eventually decompose, but it was concluded that microplastics would form

Fig.4 Biomass plastics and biodegradable plastics are completely different classifications. It should be noted that non-degradable materials, even if derived from biomass, accumulate in the environment

and accumulate in the environment over a long period of time. Therefore, process conversion of petroleum-derived plastics to biomass-derived plastics will not contribute to solving the marine plastic waste problem. At the end of this chapter, I would like to introduce the following four items as next-generation plastics.

2 Innovative Natural Polymers Used as Plastics

2.1 Polysaccharide-Derived

Cellulose, which is the most abundant polysaccharide, is considered to be particularly useful as a next-generation plastic, especially because of its availability and processability. For example, cellulose acetate has been particularly widely used as a substitute for the above nitrocellulose. Tobacco filters are widely recognized as microplastics in marine debris, and their degradation takes more than 10 years in the case of high acetylation. To make matters worse, the environmental release of tobacco derived nicotine and tar from this trash is a serious problem. Since cigarette filters do not require this level of durability, it is thought that the problem can be alleviated if they can be used as filters while maintaining the original biodegradability of cellulose by reducing the degree of acetylation. In addition, cellulose acetate has various applications such as packaging containers, fibers, films for protecting liquid crystal displays, and raw materials for cosmetics, and it is important to try to improve its biodegradability [14].

As described above, there are many issues to be addressed by cellulose derivatives, and as one of the next-generation plastics to be introduced, research on the development of transparent resins is summarized in another chapter. In addition, there is research on the use of starch, an isomer of cellulose, as a plastic.

A research group led by Prof. Uyama et al. (Osaka University) has developed a sheet that does not dissolve (collapse) in water by adding cellulose nanofibers to processed starch and developing a composite technology. Although starch is an inexpensive material that is readily available, it has not been actively used as a raw material for plastics due to problems such as water resistance. However, by generating a strong interaction between polysaccharides and cellulose, water resistance was improved and a transparent and high-strength sheet was formed. The mechanical strength of this sheet was more than twice that of general-purpose plastics (about 120 MPa) (Fig. 5). Moreover, after one month of exposure to the marine environment, decomposition progressed, and it became clear that the sheet showed marine biodegradability, as confirmed by the adhesion of bacteria and perforation of the sheet. The research was then expanded to the blending of thermoplastic starch and biodegradable plastic [15].

Starch is academically called α-1,4-glucan. On the other hand, Prof. Iwata and his colleagues (University of Tokyo) succeeded in synthesizing α-1,3-glucan, a linear polysaccharide with an extremely unusual structure, using the enzyme (α-1,3-glucan synthase: GtfJ) produced by dental caries bacteria during plaque formation, from inexpensive sucrose in a one-pot, in vitro, aqueous system. Unlike starch, α-1,3-glucan is insoluble in water, and the product was easily recovered without precipitation using organic solvents. The molecular weight of the polysaccharide α-1,3-glucan was increased to more than 700,000 and it could be chemically modified to form films and fibers as well as thermoplastics (Fig. 6) [16].

2.2 Protein-Derived

Proteins are polymers composed of linkedα-amino acids and have the potential to be used as next-generation plastics, as they are also called nylon 2 because of their structure. For example, there are attempts to utilize unused and waste keratin contained in wool, feathers and human hair as a fiber material (Fig. 7). Tracing back to research reports, there are older references to chemically cleave the disulfide cross-links in keratin to solubilize the entire cortex [17].

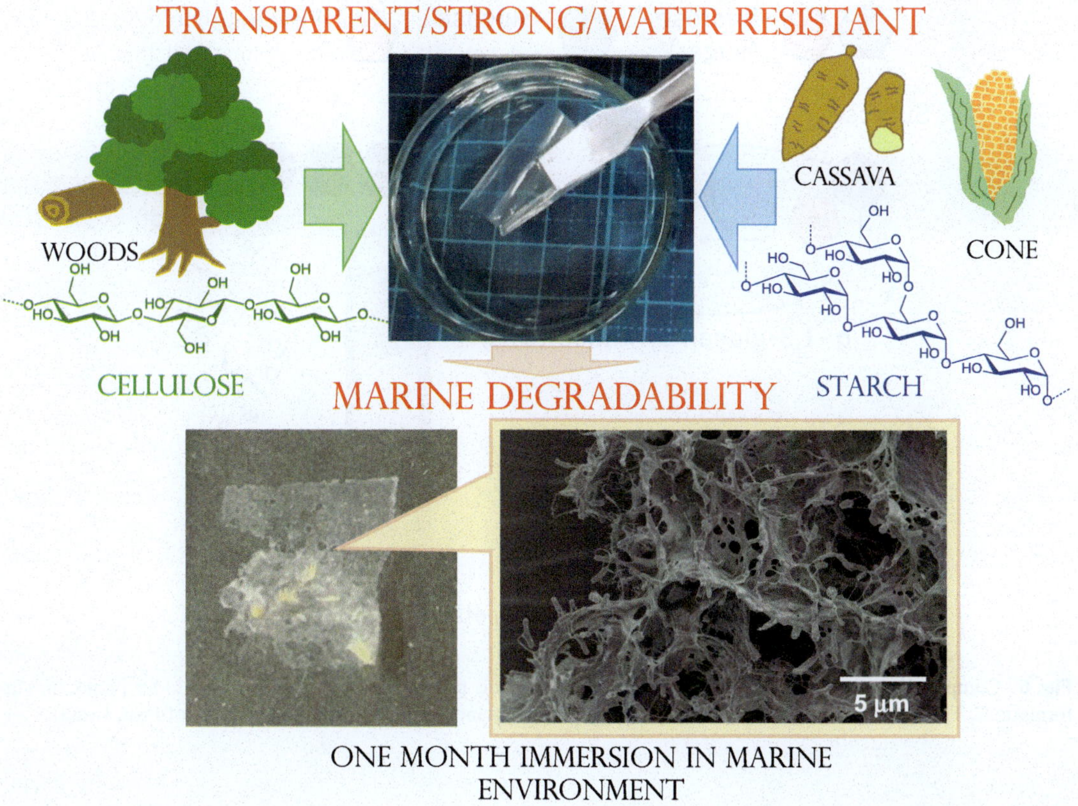

Fig. 5 Development of transparent, high-strength, marine environmentally degradable starch/cellulose composite sheet (courtesy of Prof. Uyama et al.)

Since then, various technological innovations have been accumulated and attempts have been made to use it as a biomedical material such as a cell culture scaffold material. In particular, it has been reported that biomaterials derived from human hair keratin are neuroinductive, which is interesting as a polymeric material used in the medical field [18].

Human hair is a sustainable resource that can be supplied as long as there are humans, and if it becomes widely used as a fiber or film, it will be highly significant as a next-generation plastic. The Aichi Industrial Science and Technology Center reported that they found an industrially superior method of solubilizing and concentrating keratin filaments from wool, and succeeded in giving the reduced salted keratin solution towing properties and viscosity, thereby obtaining regenerated keratin fibers. It is expected that such a technology will be widely used in the future [19].

There has been research on collagen processing for a long time. It is well known as an additive for basic cosmetics, but it is contained in all parts of livestock. If it can be extracted from waste products from processed food factories, it can be positioned as an important raw material for next-generation plastics [20].

However, collagen molecules form a triple-helical matrix in which three chains are stabilized through intra- and inter-chain hydrogen bonds, and are extremely difficult to process as they are. This is gelatin, which is widely used as a gelling agent. It has been reported that gelatin can be made into films and exhibits high mechanical strength, transparency, and barrier properties suitable for food packaging [21].

Another promising raw material is a protein called fibroin, which is the main component of silk produced by silkworms, and there is a research report that it has been made into a film [22].

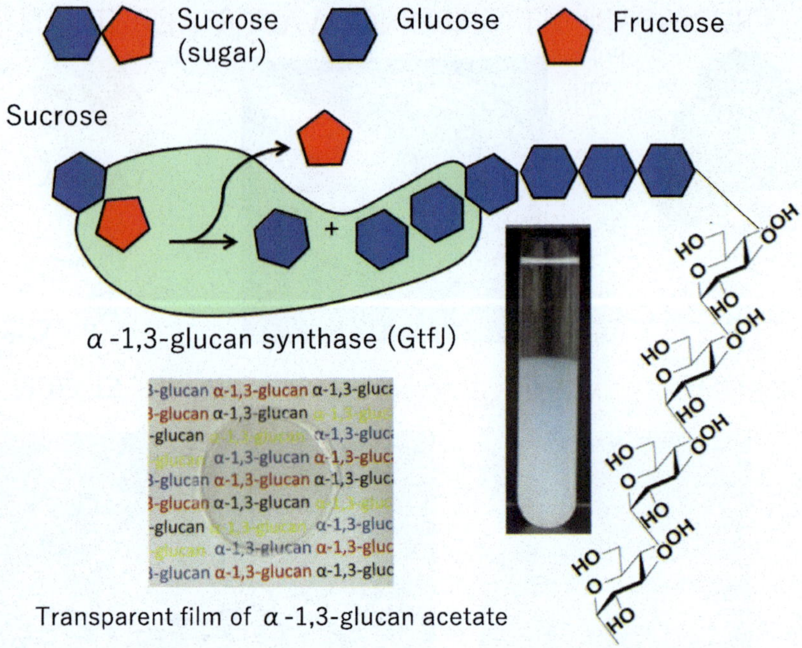

Fig. 6 Completely linear α-1,3-glucan obtained by enzymatic polymerization of sucrose (middle photo: in vitro formation). The lower left is a transparent film obtained by acetylation of α-1,3-glucan (courtesy of Prof. Iwata)

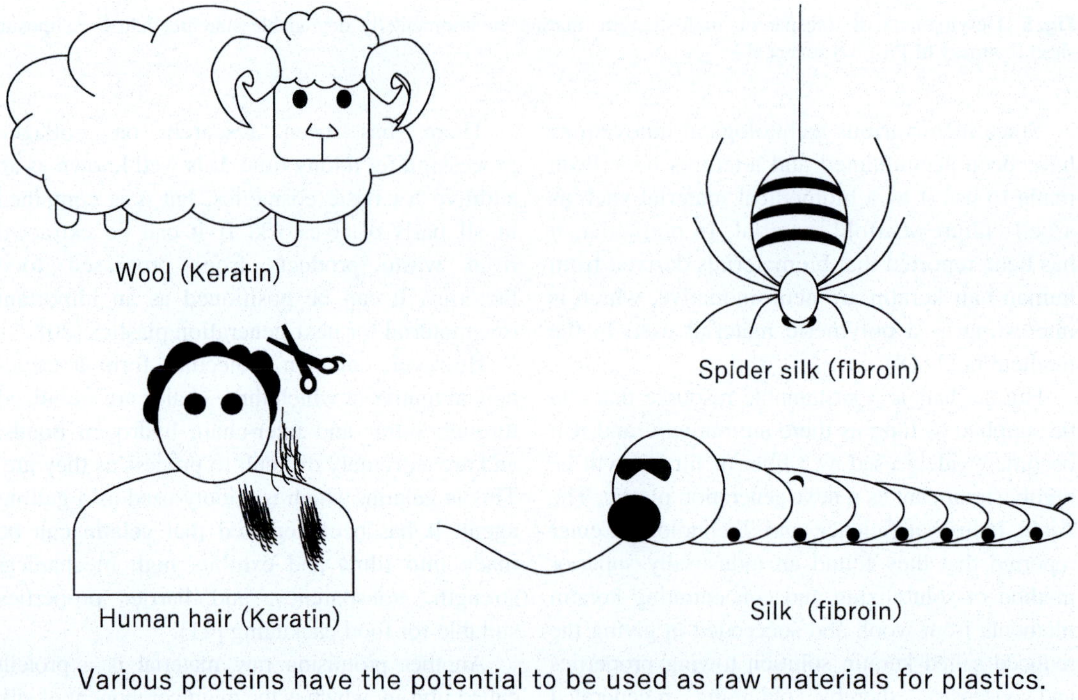

Fig. 7 Protein is also called nylon 2 structurally, and is attracting attention as a recyclable plastic because it can be processed into yarn and film in some cases

Silk was first produced in China around 6,000 years ago, and is an important recyclable raw material with good availability, as silk fabrics are now a major industry in Asian countries. There is also a recent report that silk fibroin from spider silk can be made into a film with high strength, and considering the wide variety of silkworms and spiders, it can be an excellent next-generation plastic with a wide range of properties [23].

2.3 Nucleic Acid-Derived

DNA, protein, and polysaccharide are listed as the three major biomacromolecules. We have already described above the potential and usefulness of proteins and polysaccharides as organic materials. As for the remaining DNA, it has high potential as a material because it is contained in all living organisms, although the amount of its existence is low. In fact, in Japan, the Ogata Institute for Materials Science in Chitose University of Science and Technology has extracted DNA from discarded salmon milt and is continuing active research on its use as a material. The film is provided to domestic and foreign countries and researched mainly in the field of biomedical science [24].

Many other studies have been conducted on the materialization of materials derived from natural polymers. All of them are expected to play an active role as next-generation plastics by overcoming the problems of cost and water absorbency (Fig. 8).

(1–4) Polyester as a Natural Polymer

Polyhydroxyalkanoate (PHA) is a polyester produced by hydrogen-oxidizing bacteria, discovered by Lemoigne at the Louis Pasteur Institute in France in 1926. This polyester is a carbon and energy storage material that microorganisms are equipped with when they are oligotrophic. Because of its biodegradability and biocompatibility, PHA has attracted much attention as an alternative material to petroleum-based plastics [10]. When PHA was first discovered, the homopolymer of poly (3-hydroxybutyrate) was the target material, but its crystallinity was too high, and its processability and brittleness prevented its practical use. In particular, a type of PHA called poly(3-hydroxybutyrate-*co*-3-hydroxyhexanoate) (PHBH)

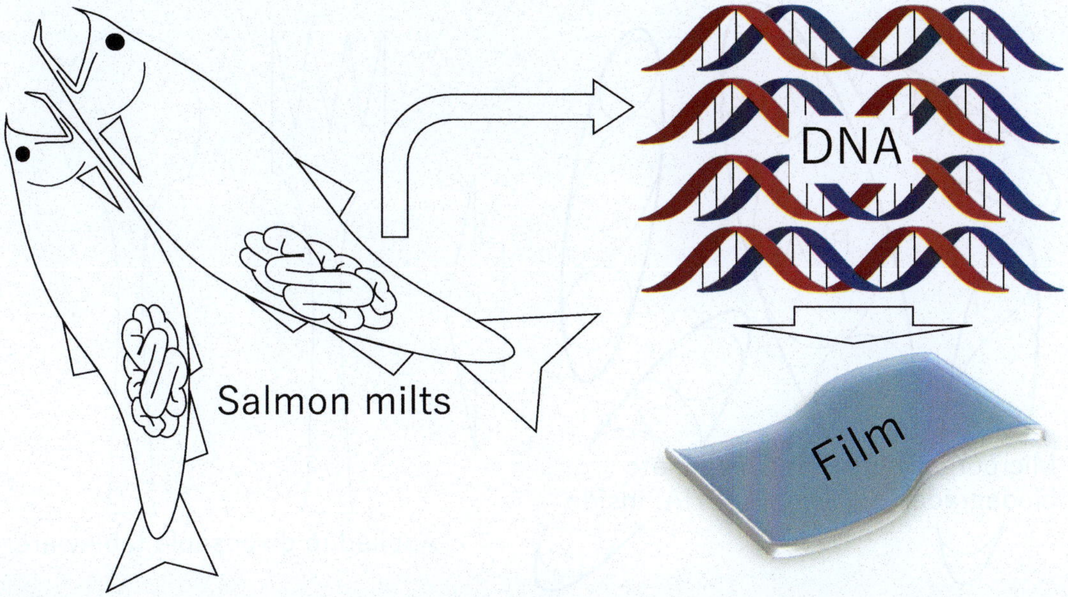

Fig. 8 Salmon milt is often discarded, and there is research to utilize the extracted DNA as a plastic film to add value to it

has been commercialized by Kaneka Corporation under the product name Green Planet™ and is expected to be used as a marine biodegradable plastic (Fig. 9) [25].

3 Biomass Plastic Showing Biodegradability

While the biodegradability of materials made mainly from natural polymers is secured by themselves without investigation, the biodegradability of polymers obtained by polymerization of biomass-derived monomers is not known without investigation. Although the structure of such polymers is limited to the monomer structure, they are more controllable than the natural polymers introduced in 1), and the structure and physical properties can be controlled to a certain degree. In this sense, as long as biodegradability and safety are ensured, it is expected to be an environmentally recyclable plastic with the necessary performance for productivity, cost, and other applications. The following are examples of social implementation. Technologies that are expected to be implemented in society in the future and plastics that are expected to become important in the near future will be introduced in a separate chapter.

3.1 Poly(Lactic Acid)

This is a typical example of biomass plastic showing biodegradability. In large-scale production, lactic acid is produced from biomass such as corn and cassava by lactic acid fermentation, and high molecular weight poly(lactic acid), PLA, is obtained by ring-opening polymerization of lactide, which is formed as a dimer of lactic acid. There are many problems in social implementation of PLA, such as crystallization speed, and various additives have been developed to overcome these problems. By adding these additives appropriately, PLA has been utilized for disposable containers, tableware, films, sheets, fibers, and non-woven fabrics. Furthermore, various molded products can be obtained by the injection method, 3D printer method, and other methods. One of the most unique applications

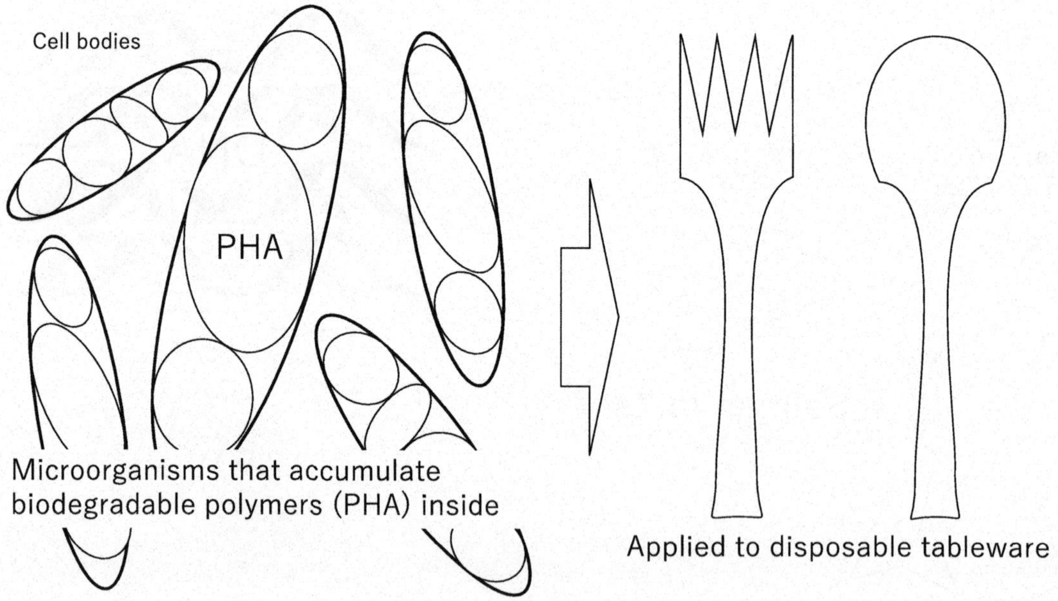

Fig. 9 Biodegradable plastics (forks, spoons, etc.) are produced from PHA, which is stored as an energy storage material in microorganisms

is as a soil degradable solid to fix cracks in reservoir rocks fractured during shale oil production for a certain period of time (Figs. 1, 2, 3, 4, 5, 6, 7, 8, 9 and 10).

PLA resin can be composted, and regulations in Europe from 2020 are stimulating the use of PLA, making it an important plastic for the next generation. Currently, NatureWorks (U.S.A.), the largest producer of PLA, produces more than 150,000 tons, and more than 15 other companies around the world produce PLA or PLA-containing resins.

3.2 Poly(Glycolic Acid)

Poly(glycolic acid) (PGA) is one of the simplest aliphatic polyesters without the methyl group in the side chain of PLA, and has been known since 1945 as a representative biodegradable plastic. On the other hand, its high mechanical strength and the water-solubility of its monomer, glycolic acid, led Davis & Geck to develop a biodegradable suture called Dexon in 1962 [26]. Sutures made of PGA are basically positioned as a type of absorbable suture that do not require suture removal. However, since it is necessary to maintain the strength for a necessary time, sutures whose degradation rate is adjusted by copolymerization with PLA, for example, are now being used and are being evaluated as a material for biomedical engineering. It is also widely used in indwelling (implantable) medical devices, and there are examples of application to anastomotic rings, plates, pins, rods, and screws (Fig. 11).

PGA, on the other hand, has an extremely high gas barrier property that is about 100 times higher than PET and about 1000 times higher than PLA. PGA can significantly improve the gas barrier performance of PET bottles when it is multilayered with PET resin. PGA is also used for nonwoven fabrics, and there are examples of its use in filters and sanitary products, for example, to control bad odors such as ammonia and amine compounds. As with PLA, there is a possibility that the application field will expand from the so far limited medical field to a much wider one, such as the growing use of PLA as a material for shale oil drilling, mainly in U.S.A [27].

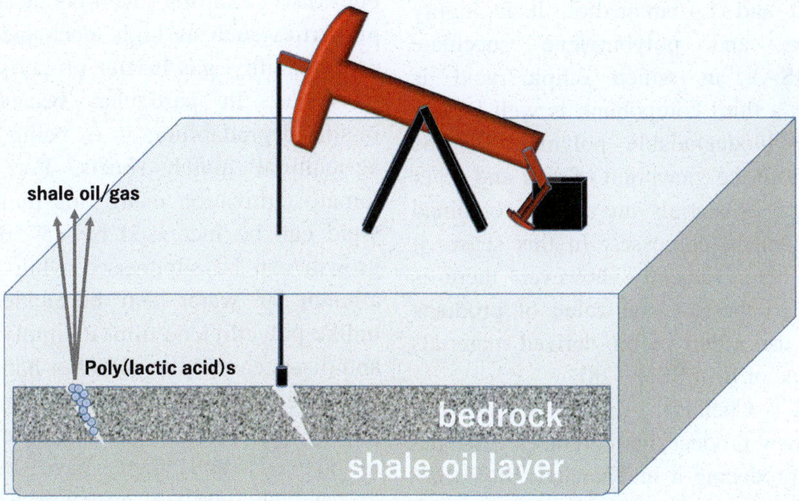

Fig. 10 Fracture closure is prevented by injecting polyl(actic acid) particles into the fractures that form when the bedrock layer above the shale oil layer is fractured. This allows oil and gas to be obtained efficiently

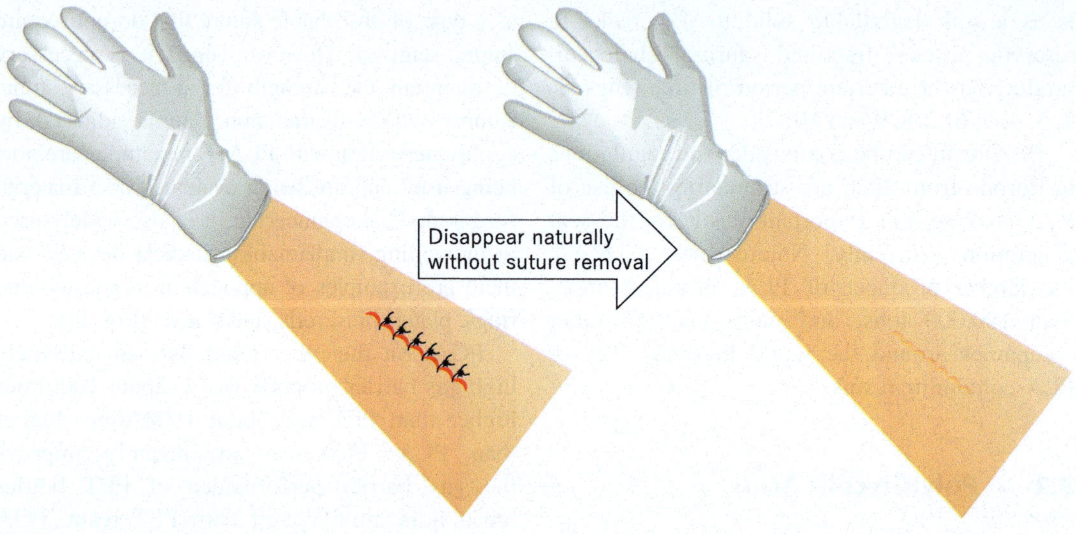

Fig. 11 Biodegradable poly(glycolic acid) derivatives are utilized as medical sutures that do not require tooth extraction

4 Petroleum-Derived Plastics Showing Biodegradability

4.1 Poly(Butylene Succinate)

Polybutylene succinate (PBS) is an aliphatic polyester obtained by polycondensation of succinic acid and 1,4-butanediol. It is highly biodegradable, and polybutylene succinate adipate (PBSA), in which adipic acid is introduced as a third component, is well known as a marine biodegradable polymer. On the other hand, from the viewpoint of cost and other factors, the raw materials are mainly obtained from petrochemical processes. In this sense, it is classified here. Recently, however, there is a movement to increase the value of products by daring to use a part of bio-derived materials under the name of BioPBS™ [28].

Mitsubishi Chemical Corporation has developed a new product that combines PBS and PLA, and is producing it in Thailand. BioPBS™ was started from the use for agricultural mulch film, but since 2019, the most famous applications, straws and paper cups, have been promoted by the Keikyu Group and the Washington Hotel.

4.2 Poly(Butylene Adipate Terephthalate)

Poly(butylene adipate terephthalate) (PBAT) is an aliphatic polyester with aromatic terephthalic acid randomly introduced into it. PBAT is not only biodegradable but also exhibits well-balanced functional properties such as high mechanical properties, processability, gas barrier property, and solvent resistance. In particular, because it shows in-soil degradability, it is being expanded to agricultural mulch sheets. For example, in tomato cultivation using this mulch sheet, the yield can be increased by 15–50% and weed growth can be suppressed while reducing the amount of water and herbicides. Moreover, unlike polyethylene film, it simply breaks down and disappears in the soil after harvest, reducing labor and costs. The introduction of aromatic groups improves the performance of polymers in general, and there have been attempts to actively utilize aromatic biomolecules. The authors have paid particular attention to cinnamic acids and developed a series of biopolyesters [29] (Fig. 12).

Fig. 12 Application example of mulch sheet for agriculture. The black part is mulch sheet, which can suppress weed growth while reducing the amount of water and herbicides. By using soil degradable PBAT, it can be proved by simply ploughing it into the soil

5 Biomass Plastic Showing Biodegradability Under Special Conditions

5.1 Bio-PET

PET is obtained by converting paraxylene and ethylene, which are originally obtained from fossil resources, into the monomers terephthalic acid and ethylene glycol, respectively, by a chemical process, and then polymerizing these monomers. On the other hand, ethylene glycol can be easily obtained from ethanol produced by fermentation, and bio-PET has been produced in India since around 1990. In recent years, PET 30 made from bio-ethylene glycol and petroleum-derived terephthalic acid has been distributed ahead of other PET resins due to the social background of promoting the shift away from petroleum. The production base of PET 100 was established by using terephthalic acid obtained from bio-paraxylene by Toray Industries, Inc. and Gevo [30]. Inherently, those that have the potential to revert to monomers by hydrolysis and the monomers are natural products may be able to function as biodegradable plastics in the future, if the conditions are right. Of course, if the degradation time is too long, it may lead to the destruction of the environment and ecosystem, which is out of the question (Fig. 13).

(4–2) Biopolyamides

Some of polyamides are produced from castor oil extracted from castor bean seeds. Castor oil does not compete with the human and animal food chain because it grows in arid areas that are unsuitable for cultivation. The main component of castor oil is ricinoleic triglyceride, which is thermally decomposed into its 11- and 7-carbon components. Chemical conversion of the 11-carbon component yields 11-aminoundecanoic acid, an omega-amino acid, which can be polymerized to yield polyamide 11 [31] (Fig. 14). Organico, the predecessor of Arkema, began selling this product in 1947, and it can be said to be the first biopolyamide. On the other hand, if castor oil is heated and melt-cracked in a concentrated alkaline solution, it can be split into 10-carbon and 8-carbon components. From the 10-carbon component, sebacic acid can be chemically derived and further converted to 1,10-decanediamine, yielding a series of biopolyamides such as polyamides 1010, 1012, 10 T (T is terephthalic acid), 610, and 410. Polyamides have long been used as engineering plastics because they have amide bonds that are far more stable than esters, and because of their hydrogen bonding properties, they have strong intermolecular forces and excellent thermomechanical performance. The properties of plastics differ depending on the carbon number, so they are used in a wide range of applications. Adipic acid is obtained by hydrogenation of cis, cis-muconic acid produced by microorganisms, and adipic acid can be converted to hexamethylenediamine, then polyamide 66 produced from these materials can also be used as a biomass plastic. Polyamide 66 produced from these raw materials can also be obtained as a biomass plastic. Furthermore, it is reported that polyamide 66 and Nylon6 can also be decomposed by white rot fungi, and degrading enzyme of its oligomers has also been identified (Fig. 13).

In addition, polyamide 56 is also synthesized because decarboxylation of the amino acid L-lysine yields pentamethylenediamine. The degradation rate of polyamide 56 is reported to be faster than that of polyamide 66 [32]. Another biopolyamide, polyamide 4, is derived from gamma-aminobutyric acid (GABA), a component widely possessed by living organisms, and is a next-generation nylon that also exhibits marine biodegradability. In general, the smaller the carbon number, the more biodegradable the material is. This is understandable, considering that proteins are biodegradable and structurally can be regarded as derivatives of polyamide 2. The authors have developed a series of biopolyamides using itaconic acid, an unsaturated aliphatic dicarboxylic acid with five carbons produced by *Aspergillus* sp, as a starting material. This is a biomass plastic and a next-generation degradable plastic that becomes biodegradable only after disposal. Details of this are described in Chap. 2.

5.2 Bio-Polyurethane

Polyurethanes (PU) are obtained from a raw material called polyol with multiple hydroxyl groups and isocyanate. Bio-PU is obtained by polymerization of polyol derived from plant oils such as soybean oil or castor oil with isocyanate. Generally, the biobase degree of bio-PU foam is about 15%, but Uyama Laboratory of Osaka

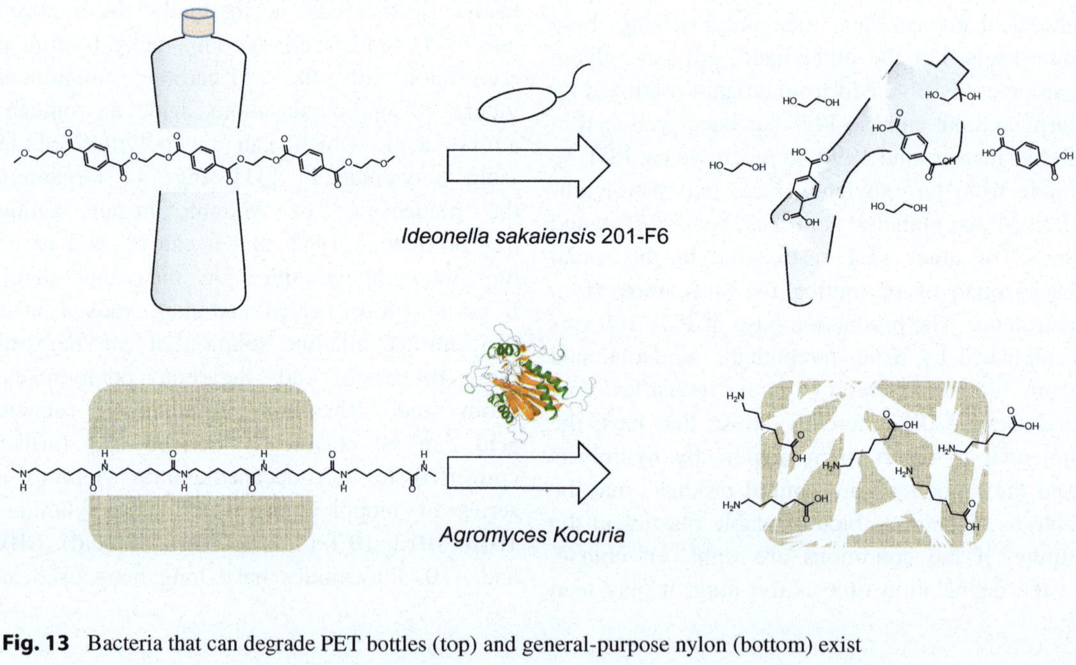

Fig. 13 Bacteria that can degrade PET bottles (top) and general-purpose nylon (bottom) exist

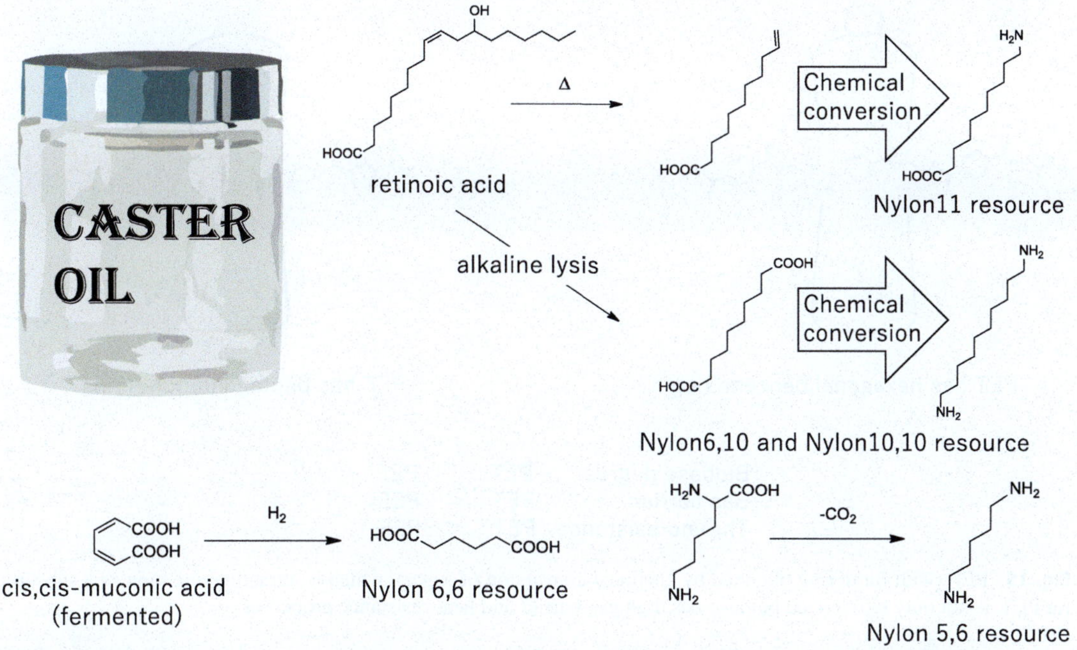

Fig. 14 Raw materials for castor oil derived from castor bean and nylon derived from biomolecules derived from microorganisms

University synthesized hyperbranched PLA with a large amount of hydroxyl groups and reacted it with isocyanate to obtain bio-PU with a biobase degree of 50%. In addition, various bio-PUs have been developed by domestic and overseas companies such as Mitsui Chemicals, Toyo Soflan Tech, Dainichi Seika, Cargill, and Covesto.

(4–4) Poly(ethylene furanoate)

Poly(ethylene furanoate) (PEF) has been developed as a substitute for PET and is obtained by polymerizing furan dicarboxylic acid (FDCA) produced from bio-derived sugars with bio-derived ethylene glycol. The furan ring is aromatic and PEF has physical properties similar to PET, but has higher gas barrier properties than PET. Toyobo is producing and marketing products that take advantage of these characteristics. In addition, academic research is being developed at Hokkaido University, Portsmouth University, and Tokyo Institute of Technology [33] (Fig. 15).

PEF is superior to PET.

This book describes the photo-switched biodegradability in plastics that have been developed based on a clear bonding mode of degradation such as hydrolysis from biomolecules, and then introduces photo-switchable biodegradable plastics to be developed in the future as next-generation plastics. Actually it takes a long time to complete the works with such a wonderful topic and still on the way. Here we put together the elemental technologies to achieve this goal in each chapter.

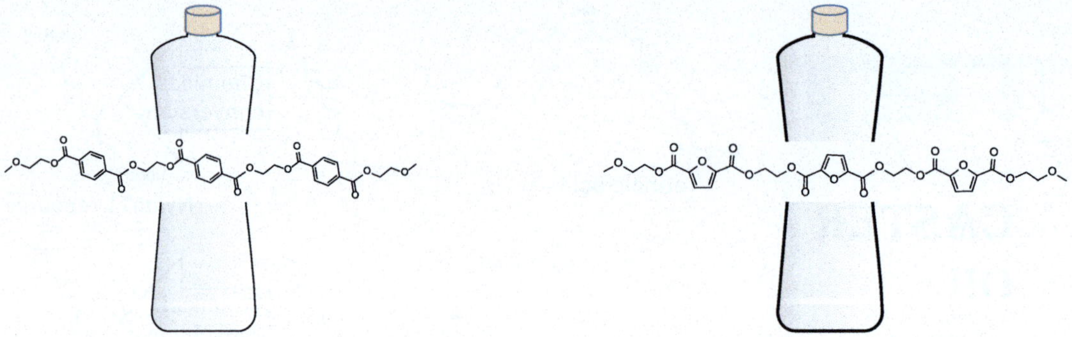

PET has hexagonal benzene ring

PEF has pentagonal furan ring

Biobase degree	PET	<	PEF
Bas barrier	PET	<	PEF
Thermoresistance	PET	<	PEF

Fig. 15 PEF (with furan ring obtained by chemical conversion of sugar), which is attracting attention as a substitute for PET, is not only bio-derived but also has high gas-barrier and heat-resistance properties

References

1. S. Miller, A production of amino acids under possible primitive earth conditions. Science **117**, 528 (1953)
2. E. Nakamura et al., On the origin and evolution of the asteroid Ryugu: a comprehensive geochemical perspective. Proc. Jpn. Acad. Ser B. **98**, 227 (2022)
3. J.A. Baross, S.E. Hoffman, Submarine hydrothermal vents and associated gradient environments as sites for the origin and evolution of life. Orig. Life **15**, 327 (1985)
4. M. Goodman, D.A. Tagle, D.H.A. Fitch, W. Bailey, J. Czelusniak, B.F. Koop, P. Benson, J.L. Slightom, Primate evolution at the DNA level and a classification of hominoids. J. Mol. Evol. **30**, 260 (1990)
5. G. Williams, Shellac as musical plastic. J. Am. Musicol. Soc. **74**, 463 (2021)
6. S.C. Rasmussen, From parkesine to celluloid: the birth of organic plastics **60**, 8012 (2021)
7. C.B. Hollabaugh, L.H. Burt, A.P. Walsh, Carboxymethylcellulose. Uses and applications. Ind. Eng. Chem. **37**, 943 (1945)
8. J. Han, Y. Guo, H. Wang, K. Zhang, D. Yang, Sustainable bioplastic made from biomass DNA and ionomers. J. Am. Chem. Soc. **143**, 19486 (2021)
9. E. Sutermeister, F.L. Browne, *Casein and Its Industrial Applications* (Reinhold Publishing Corp, NY, 1939)
10. M. Lemoigne, Production of β–hydroxybutyric acid by certain bacteria of the B. subtilis group. Ann. Inst. Pasteur **39**, 144 (1925)
11. W.H. Carothers, G.L. Dorough, F.J. van Natta, Studies of polymerization and ring formation. X. The reversible polymerization of six-membered cyclic esters. J. Am. Chem. Soc. **54**, 761 (1932)
12. http://www.jbpaweb.net/english/e-bp/
13. https://www.astm.org/d6866-22.html
14. https://prtimes.jp/main/html/rd/p/000000009.000035577.html
15. R. Soni, Y. -I Hsu, T. Asoh, H. Uyama, Synergistic effect of hemiacetal crosslinking and crystallinity on wet strength of cellulose nanofiber-reinforced starch films. Food Hydrocoll. **120**, 106956 (2021)
16. S. Puanglek, S. Kimura, Y. Enomoto-Rogers, T. Kabe, M. Yoshida, M. Wada, T. Iwata, In vitro synthesis of linear α-1,3-glucan and chemical modification to ester derivatives exhibiting outstanding thermal properties. Sci. Rep. **6**, 30479 (2016)
17. W.G. Crewther, R.D.B. Fraser, F.G. Lennox, H. Lindley, The chemistry of keratins, in *Advances in Protein Chemistry*, ed. by C.B. Anfinsen Jr, M.L. Anson, J.T. Edsall, F.M. Richards (Academic Press, New York, 1965), pp. 191–346
18. P. Sierpinski, J. Garrett, J. Ma, P. Apel, D. Klorig, T. Smith, L.A. Koman, A. Atala, M.V. Dyke, The use of keratin biomaterials derived from human hair for the promotion of rapid regeneration of peripheral nerves. Biomaterials **29**, 118 (2008)
19. K. Kanayama, H. Nakanishi, Wool Keratin Refiberation Technology, Aichi Prefectural Institute of Industrial Technology Research Report 126 (2009)
20. D. Khodaei, C. Álvarez, A.M. Mullen, Biodegradable packaging materials from animal

processing co-products and wastes: an overview. Polymers **13**, 2561 (2021)
21. LWT, Volume 117, 2020, 108617 Preparation and characterization of blended edible films manufactured using gelatin, tragacanth gum and, Persian gum
22. Y. Kawahara, K. Furukawa, T. Yamamoto, M. Masuda, T. Furuzono, Development of flexible silk fibroin film. J. Silk Soc. Jpn. **15**, 3–6 (2006)
23. K. Tsuchiya, T. Ishii, H. Masunaga, K. Numata, Spider dragline silk composite films doped with linear and telechelic polyalanine: effect of polyalanine on the structure and mechanical properties. Sci. Rep. **8**, 3654 (2018)
24. C. Ceron Jayme, L. Barcelosde Paula, N. Rezende, I. Rodrigo Calori, L. Pereira Franchi, A. Claudio Tedesco, DNA polymeric films as a support for cell growth as a new material for regenerative medicine: compatibility and applicability. Exp. Cell Res. **360** (2), 404–412 (2017)
25. https://www.kaneka.co.jp/solutions/phbh/
26. D.K. Gilding, A.M. Reed, Biodegradable polymers for use in surgery—polyglycolic/poly (lactic acid) homo- and copolymers: 1
27. https://www.kuredux.com/about/index.html
28. https://www.mcpp-global.com/fileadmin/mcpp_data/documents/Products/4-Bio/BioPBS_-_AW_PTTMCC_Brochure_2019_A5_.pdf
29. https://plastics-rubber.basf.com/global/en/performance_polymers/products/ecoflex.html
30. I.T. Takehana, H. Yamaji, Y. Maeda, K. Toyohara, K. Miyamoto, Y. Kimura, K. Oda, A bacterium that degrades and assimilates poly(ethylene erephthalate). Science **351**(6278), 1196–1199 (2016)
31. https://hpp.arkema.com/en/product-families/rilsan-polyamide-11-resins/
32. https://projectdb.jst.go.jp/grant/JST-PROJECT-12102330/
33. D. James Gordon Napier, L. James, Polyesters from heterocyclic components, US 2551731 A (1951)

Biorefinery: Microbial Production of Building Blocks from Plant Resources for the Synthesis of Bioplastics

Hideo Kawaguchi, Takashi Sazuka and Dao Duy Hanh

1 Biorefinery to Utilize Lignocellulosic Biomass

1.1 Biorefinery Feedstocks

Plant biomass is a renewable resource for energy and chemicals since the organic materials are started from carbon dioxide (CO_2) fixation by photosynthesis. In 1990s, edible plant biomass, such as glucose or starch, was used for the production of bioethanol as an alternative to gasoline. However, the use of edible biomass competes with the production of food and feeds. To avoid the competition, inedible lignocellulose and waste, which come from woody and cop biomass and agricultural wastes, are currently considered as the next generation biomass feedstocks for biorefinery [21]. Lignocellulose is predominantly composed of is predominantly composed of cellulose (25–55%), hemicellulose (11–50%), and lignin (10–40%) [72]. The annual global production of lignocellulose accounts for about 180 giga tons, in which about 8–20 giga tons of the primary biomass remains potentially accessible [35]. The carbon of lignocellulosic biomass is exclusively derived from atmospheric CO_2 fixed by photosynthesis. Global energy-related CO_2 emission reached 33.4 giga ton in 2019 (https://www.iea.org/articles/global-energy-review-co2-emissions-in-2020). Thus, Lignocellulosic biomass has a great potential as alternative feedstocks to fossil resources for industrial production to mitigate increasing CO_2 levels in the atmosphere towards sustainable society.

In the last decades, various plant resources have been investigated as biorefinery feedstocks for the production of fuels and chemicals [15]. The utilization of agricultural wastes, such as corn stover, sugarcane bagasse and, sorghum bagasse, has been studied for the production of bioethanol and value-added chemicals. However, the cultivation of these crops potentially and closely relevant to the production of food and feed. In contrast, *Miscanthus* and switchgrass are being developed into biomass energy crops that are independent on the availability of land for food and feed production

H. Kawaguchi (✉) · D. D. Hanh
Graduate School of Science, Technology and Innovation, Kobe University, 1-1 Rokkodai, Nada, Kobe 657-8501, Japan
e-mail: kawaguchi_h@port.kobe-u.ac.jp

D. D. Hanh
e-mail: dao-duyhanh@people.kobe-u.ac.jp

H. Kawaguchi
Engineering Biology Research Center, Kobe University, 1-1 Rokkodai, Nada, Kobe 657-8501, Japan

T. Sazuka
Bioscience and Biotechnology Center, Nagoya University, Furo, Chikusa 464-8601, Nagoya, Japan
e-mail: sazuka@agr.nagoya-u.ac.jp

T. Kaneko (ed.), *Photo-switched Biodegradation of Bioplastics in Marine Environments*,
https://doi.org/10.1007/978-981-99-4354-8_2

unless these crops are cultivated in agricultural land. As woody biomass, in addition, poplar, willow, and Jatropha, which can be cultivated in marginal land, has drawn much research interest as biorefinery feedstocks. Thus, development of technologies to utilize lignocellulose derived from plants which are suitable for the cultivation in marginal land is needed to avoid the competition with food and feed production.

1.2 Sorghum: As a Leading Candidate of Biorefinery Feedstock

Sorghum (*Sorghum bicolor* (L.) Moench) is monocot, grass, *panicoideae*, and C_4 species crop, like maize (*Zea Mays*), and sugarcane (*Saccharum officinarum*). Sorghums were classified into five groups agronomically; (1) forage sorghum, (2) biomass sorghum, (3) sweet sorghum, (4) grain sorghum, (5) bloom sorghum. In a view point of biorefinery, biomass sorghum and sweet sorghum are important. Because of the superior biomass of leaves and stems, forage sorghum (*Sorghum bicolor* (L.) Moench) are widely cultivated all over the world. Recently, to obtain higher biomass for the bioethanol production, new lines are selected and bred as the biomass sorghums. Sweet sorghums are another sub-group of sorghum, which accumulate sugars like sugarcane, and cultivated widely in the last century, however, rarely cultivated commercially now. Compared with sugarcane, sorghum could be cultivated more widely; from the equator to the border between temperate zone and subarctic zones. In addition, annual productivity of biomass of sorghum is comparable with that of sugarcane, and the biomass yield was equal or greater than that of corn while using 33% less water [50]. The water-use efficiency and drought tolerance of sorghum make it adaptable to cultivation on marginal lands unsuitable for conventional agriculture. These features are advantage of sorghum as one of leading candidates for biorefinery feedstock to supply a large amount of biomass globally without competing the production of food.

Sorghum genome (var. BTx623) was firstly determined in 2005 [30], and ~34,000 genes in the 10 chromosome is annotated in the 732 Mbp of the genome (https://phytozome-next.jgi.doe.gov/info/Sbicolor_v3_1_1). Additionally, whole genome sequences of other three grain sorghums and one sweet sorghum is available (https://phytozome-next.jgi.doe.gov/info/Sbicolor_v3_1_1). Cytogenetically, sorghum is a diploid having two complete sets ($2n=2x=20$) of chromosomes whereas sugarcane is auto polyploid having ~$12\times$ of chromosomes [64]. Due to diploidy and availability of whole genome sequences, sorghum breeding is accelerated especially by DNA-marker assisted breeding [46]. There are also indispensable alleles (or genes) which have been used in classic breeding in sorghum, and these are also valuable for biorefinery. For example, the brown-midrib (bmr) mutation involved in cell wall composition, generally reduces lignin content [3, 48, 51]. Brown midrib sorghums were shown to exhibit improved sugar conversion into ethanol of their biomass compared to conventional sorghums [7, 11].

Another example is the content of soluble carbohydrates (sucrose, glucose, and fructose), which are generally referred to fermentable sugars, and these accumulate in stems of sweet sorghums like sugarcane [17]. Furthermore, genetic mechanism in heterosis is gradually becoming clearer in sorghum. Quantitative trait locus (QTL) analysis of the progeny of a high-biomass sorghum F_1 hybrid revealed that five QTLs were responsible for culm length, and the developed pyramided lines, using the five homozygous dominant alleles, produced biomasses like the original F_1 line [18]. In other words, it is developed as the novel breeding approach which can design the genome of an inbred line from the hybrid line in sorghum. Thus, sorghum have a leading candidate as biorefinery feedstocks of C_4 grass.

Dilute acid-pretreated sorghum bagasse contained cellulose (50–59%) and hemicellulose (7–20%), respectively [26]. Thus, the hydrolysate contained glucose and xylose, which were released from these polysaccharides after enzymatic hydrolyzation, can serve as carbon sources for the following fermentation (Fig. 1). The juice of sweet sorghum extracted from the stalk is predominantly sucrose with variable levels of glucose and fructose as carbon sources, and amino acids as a source of nitrogen for fermentation [24]. Thus, these two types of biomass were used for the production of biofuels and biochemicals (Table 1).

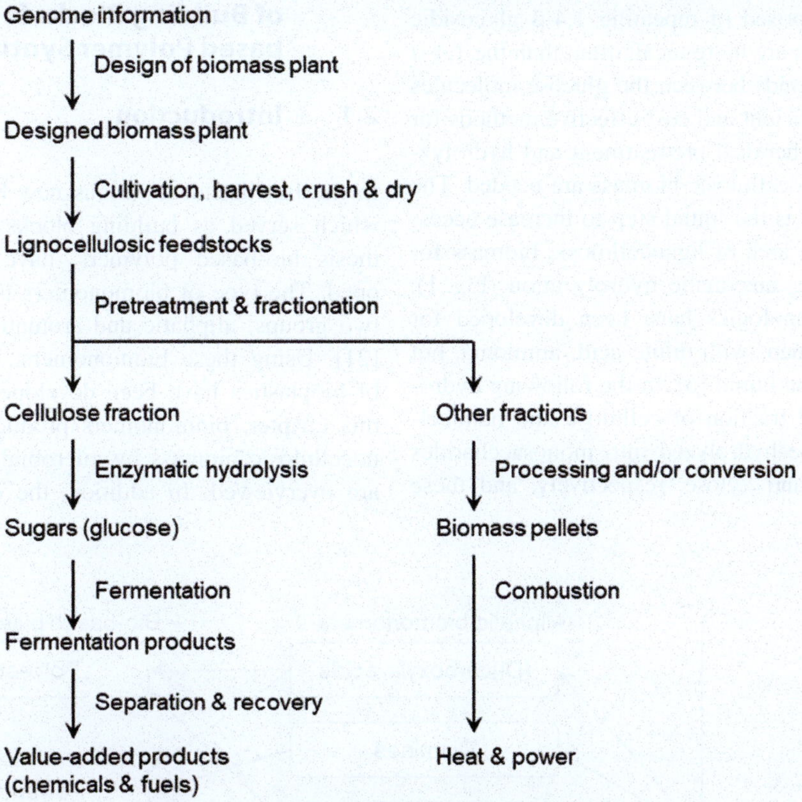

Fig. 1 Scheme of biorefinery to produce energy and chemicals from biomass plant

Table 1 Fuel and energy production from sorghum biomass

Type of biomass	Product	Fermenting microorganism	References
Bagasse	Solvent (acetone, butanol, and ethanol)	*Clostridium acetobutylicum*	[71]
	Ethanol	*Saccharomyces cerevisiae*	[41, 65]
	Phenyllactic acid	*Escherichia coli*	[26]
Juice	Ethanol	*S. cerevisiae*	[52, 55, 71]
	3-amino-4-hydroxybenzoic acid	*Corynebacterium glutamicum*	[24]
	Lactic acid	*Bacillus coagulans* and *Lactobacillus rhamnosus*	[66]

1.3 Pretreatment and Hydrolyzation of Biorefinery Feedstocks

Inedible cellulose is more recalcitrant to the hydrolyzation than edible starch. Because cellulose is comprised of repeating 1,4-β glycosidic bonds, which are more recalcitrant than the 1,4-α glycosidic bonds between the glucose molecules of starch, efficient and cost-effective methods for the physicochemical pretreatment and hydrolyzation of lignocellulosic biomass are needed. The pretreatment is the initial step to increase accessible surface area of lignocellulosic biomass for the following enzymatic hydrolyzation (Fig. 1). Various technologies have been developed for the pretreatment with dilute acid, ammonia, hot water, and ion liquid [6]. In the following hydrolyzation, the fraction of cellulose and hemicellulose can be hydrolyzed into monosaccharides of glucose and xylose, respectively, and these monosaccharides can be used as substrates for microbial fermentation for the production of fuel and chemicals [42].

2 Microbial Production of Building Blocks for Bio-based Polymer Synthesis

2.1 Introduction

In the last decades, various new biomonomers, which served as building blocks for the synthesis bio-based polymers, have been developed. The type of biomonomers is divided into two groups; aliphatic and aromatic compounds [21]. Using these biomonomers, various types of bioplastics have been developed (Fig. 2). In this chapter, biomonomers produced from lignocellulosic biomass by microbial fermentation are overviewed. In addition, the current status

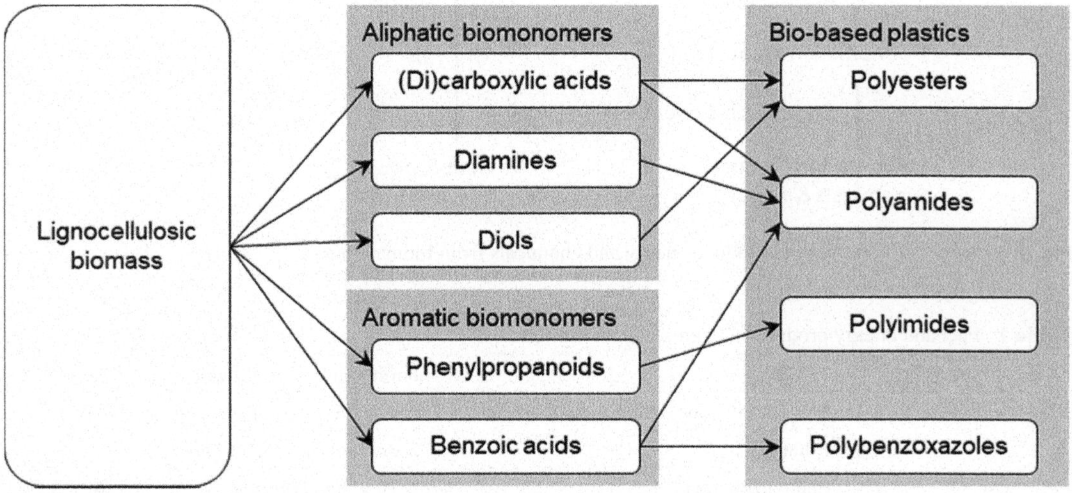

Fig. 2 Biomonomers produced from lignocellulosic biomass for the following chemical synthesis of bio-based plastics

and future perspectives of emerging bioplastics made from the biomonomers are discussed. Details of the development of bio-based polymers are excellently summarized in Chap. 1.

2.2 Production of Aliphatic Compounds from Lignocellulosic Biomass

Poly(butylene succinate) (PBS) and poly(lactic acid) (PLA) are commercially available and biodegradable polyesters. PBS is chemically synthesized from the mixture of succinic acid and 1,4-butanediol [57], whereas PLA is chemically polymerized from lactic acid through the dilactide [43]. Currently, the vast majority of succinic acid and lactic acid are produced by fermentation. Various microorganisms were used to produce lactic acid that was produced from both grassy [19, 47, 67] and woody [75] biomass (Table 2). Succinic acid was produced from the hydrolysates of corn stover [4], cassava pulp [53], and beachwood xylan [74]. In addition to succinic acid, various dicarboxylic acids such as fumaric acid [37], malic acid [38], and adipic acid [10], and *cis,cis*-muconic acid [63] were produced from lignocellulosic biomass or its model structural components (Table 2). These dicarboxylic acids can serve as building blocks to synthesize polyester and polyamide. Diamines can serve as counterparts with (di) carboxylic acid stated above to synthesize bio-based polyamides. Cadaverine (1,5-diaminopentane) is one of the most promising chemicals for the synthesis of the bio-based polyamide. However, cadaverine was produced from only edible glucose or soluble starch by recombinant *Corynebacterium glutamicum* but nothing from inedible lignocellulosic biomass [31, 58].

2.3 Production of Aromatic Compounds from Lignocellulosic Biomass

Incorporating rigid ring moieties into a polymer main chain enhances the thermomechanical performance of the polymer [62]. Thus, aromatic compounds can serve as monomers for the production of high-performance plastics, such as polybenzoxazole and polyimide, with greater mechanical strength or thermostability as compared with commodity plastics. Shikimic acid derivatives such as *p*-coumaric acid and caffeic acid are precursors for lignin biosynthesis and

Table 2 Microbial production of aliphatic biomonomers from lignocellulosic biomass

Product	Biomass	Microorganism	Titer (g/L)	References
Lactic acid	Sugarcane bagasse	*Bacillus* sp.	185	[47]
	Corn stover	*Bacillus coagulans*	97.6	[19]
	Mixture of softwood prehydrolysate and paper mill sludge	*Lactobacillus rhamnosus*	60	[75]
	Rice bran	*L. rhamnosus*	59	[67]
Succinic acid	Corn stover	*Actinobacillus succinogens*	39.6	[4]
	Cassava pulp	*Escherichia coli*	41.5	[53]
	Beachwood xylan	*E. coli*	14.4	[74]
Fumaric acid	Alkaline-pretreated corncob	*Rhizopus oryzae*	41.3	[37]
Malic acid	Corn straw	*Rhizopus delemar*	120	[38]
Adipic acid	Milled corncob	*Thermobifida fusca*	0.22	[10]
cis,cis-Muconic acid	Benzoate, *p*-coumarate, phenol, or 4-hydroxybenzoate	*Pseudomonas putida*	13.5	[63]

Table 3 Microbial production of aromatic biomonomers from lignocellulosic biomass

Product	Biomass	Microorganism	Titer (g/L)	References
Caffeic acid	Kraft pulp	*Escherichia coli*	0.233	[23]
p-Coumaric acid	Phosphoric acid swollen cellulose	*Streptomyces lividans*	0.5	[28]
Phenyllactic acid	Sorghum bagasse	*E. coli*	1.7	[26]
	Kraft pulp	*E. coli*	2.8	[27]
3-Amino-4-hydroxybenzoic acid	Kraft pulp	*Corynebacterium glutamicum*	3.1	[44]

also candidate biomonomers, but they are exclusively synthesized and accumulated in plants with a trace amount. To produce these plant-specific secondary metabolites in an industrial scale, recombinant microorganisms been developed over the past decade (Table 3). Despite the considerable efforts, the use of lignocellulosic biomass for the production of aromatic biomonomers are still limited. Caffeic acid was produced from Kraft pulp as woody biomass by a recombinant *Escherichia coli* strain expressing the genes encoding 4-hydroxyphenyllactate 3-hydroxylase (*hpaBC*) from *Pseudomonas aeruginosa* and tyrosine ammonia lyase (*fevV*) from *Streptomyces* sp. [23]. *p*-Coumaric acid, the precursor of caffeic acid in plant, was produced from phosphoric acid swollen cellulose (PASC) by recombinant *Streptomyces lividans* expressing tyrosine ammonia lyase derived from *Rhodobacter sphaeroides* and endoglucanase [28]. Phenyllactic acid (PhLA), which is widely existing in honey and fermented food, and served as aromatic biomonomer [14], was produced from both sorghum bagasse [26] and Kraft pulp [27] by recombinant *E. coli* strain expressing phenylpyruvate reductase from *Wickerhamia fluorescens*.

3-Amino-4-hydroxybenzoic acid (3,4-AHBA) is a metabolic intermediate of the biosynthesis of an antibiotics, grixazone, in *Streptomyces griseus* [56]. To produce 3,4-AHBA from sorghum biomass, *C. glutamicum* was metabolically engineered to express the responsible genes derived from *S. griseus* [22, 24]. 3,4-AHBA was also produced from lignocellulosic biomass of Kraft pulp, and the cellulose-based 3,4-AHBA was used as a precursor for ultra-thermoresistant bioplastics [45] (Table 3). Besides them, various aromatic biomonomers, such as *p*-aminobenzoic acid [33], 4-aminocinnamic acid [59], and shikimic acid [12, 32], were produced by microbial fermentation. However, these biomonomers have been produced only from edible glucose or model lignin components but not from lignocellulosic biomass. Compared to aliphatic biomonomers, titers of aromatic biomonomers are still relatively low, and use of lignocellulosic biomass for the production are limited (Tables 2 and 3). Metabolic engineering of microbial cells for improved production of aromatic biomonomers and the application for the production from lignocellulosic biomass is needed in further studies.

2.4 Production of Itaconic Acid from Lignocellulosic Biomass

Itaconic acid was selected as one of 12 platform chemicals for the production of value-added bio-based chemicals and materials [68]. Currently, itaconic acid is paid attention in various fields such as food processing, agricultural production, pharmaceutical industry, degradable polymers production and water treatment [8, 9, 49, 60]. More recently, bio-based itaconic acid served as a precursor to synthesize a novel bio-nylon with enzymatic digestibility [2]. It is estimated that global production of itaconic acid will exceed 41 giga ton/year (https://docplayer.net/95007975-Wp-8-1-determination-of-market-potential-for-selected-platform-chemicals.html) and the market value is forecasted to be 102.3 million USD by 2022. To avoid the competition with the production of food and feeds, the use of inedible lignocellulosic biomass as alternative carbon source to edible glucose and starch is being extensively studied.

Due to multiple steps and low productivity in chemical reaction for itaconic acid synthesis [40], itaconic acid is commercially produced by fermentation. A filamentous fungus *Aspergillus terreus* produced itaconic acid from edible starch or glucose by with the titer of 160 g/L and the yield of 0.6 g/g of glucose [34], and the production from inedible lignocellulosic biomass is extensively studied. The efforts of using lignocellulosic biomass as substrate aim to reduce the cost of itaconic acid production from 3.0 US$/kg to 1.5 US$/kg, so that it can be competitive with petrochemicals [8]. However, *A. terreus* is sensitive to fermentation inhibitors contained in hydrolysates of lignocellulosic biomass. To overcome the limitation, much efforts have been paid to develop a bioprocess to produce itaconic acid from lignocellulosic biomass.

In many research studies, *A. terreus* has been used as a model itaconic acid producer. An enzymatic hydrolysate was prepared from steam-exploded corn stover to investigate the production of itaconic acid. A mutant strains of *A. terreus* produced itaconic acid at the concentration of 19.3 g/L, whereas the wild-type *A. terreus* failed to grow in the undetoxified enzymatic hydrolysate [36] (Table 4), suggesting that the use of mutant strain with tolerance to the fermentation inhibitors is effective for itaconic acid production from lignocellulosic biomass. Another approach is detoxification of hydrolysate of lignocellulosic biomass. With wild-type *A. terreus*, itaconic acid was produced from organosolv-pretreated beech wood cellulose (7.2 g/L), in which most of fermentation inhibitors were removed with organic solvent [61]. Additionally, detoxified enzymatic hydrolysate of corn stover was used as a substrate for itaconic acid production by wild-type *A. terreus* (33.6 g/L), in which acetic acid contained in the hydrolysate was removed with activated charcoal to attenuate fermentation inhibition [39]. With detoxified hydrolysis of wheat bran, *A. terreus* demonstrates the highest concentration of itaconic acid (49.7 g/L) from lignocellulosic biomass [69]. As woody biomass, bleached pulp derived from *Eucalyptus* wood was also

Table 4 Itaconic acid from lignocellulosic biomass

Organisms	Biomass	Titer (g/L)	Yield (g/g)	References
Aspergillus terreus	Detoxified organosolv pretreated beech wood	7.2	0.3	[61]
	Corn stover hydrolysate	19.3	0.36	[36]
	Detoxified hydrolysate of wheat bran	49.7	0.55	[69]
	Detoxified corn stover hydrolysate activated with charcoal	33.6	0.56	[39]
	Bleached cellulose pulp hydrolysate	27.6	0.52	[29]
	Bamboo residues hydrolysate	41.5	Not applicable	[70]
Neurospora crassa	Avicel	0.02	Not applicable	[73]
Ustilago maydis co-culture with Trichoderma reesei	Cellulose	33.8	0.16	[54]
Pseudomonas putida	Lignin	1.4	0.79	[13]

used as a substrate of itaconic acid production by *A. terreus* (14.8 g/L) [29]. In addition, undetoxified hydrolyzed bamboo residue was used to evaluate itaconic acid production under different fermentation conditions. Citrate buffer was usually used in enzymatic hydrolyzation of lignocellulosic biomass since cellulase activity was increased under acidic conditions. To avoid fermentation inhibition of citrate contained in the enzymatic hydrolysate, water was used instead of citrate buffer during enzymatic hydrolyzation, and improved production of itaconic acid was observed from the resulting hydrolysate of bamboo residues without detoxification in fed-batch fermentation using *A. terreus* (41.5 g/L) [70].

Different microorganisms were examined for itaconic acid production from lignocellulosic biomass to overcome fermentation inhibition observed in *A. terreus*. However, the titer of itaconic acid in culture with *Neurospora crassa*, *Pseudomonas putida*, or a coculture of *Ustilago maydis* and *Trichoderma reesei* was still lower than that using *A. terreus* [13, 54, 73] (Table 4). Although lignocellulosic biomass has been used as alternative carbon substrate for itaconic acid production, the productivity was still lower than that from glucose. For all microorganisms, metabolic engineering is needed to control the carbon availability for itaconic acid production rather than other cell purposes. In addition, the development of process to consolidate multiple steps and to avoid fermentation inhibition is needed to improve itaconic acid production from lignocellulosic biomass.

2.5 Bioplastics Produced from Biomass

Bioplastics are alternatives to commercially available plastics made from petroleum-based raw materials, which are made from renewable biological resources, such as biomass, or are biodegradable polymers including petroleum-based plastics. Polyhydroxyalkanoates (PHAs) is a biodegradable and naturally occurring biopolyester, which is accumulated in the microbial cells of *Alcaligenes latus*, *Azospirillum rubrum*, *Azotobacter vinelandii*, *Bacillus megaterium* and *Bacillus cereus*, *Enterobacter* sp., *Leptothrix* sp., *Methylocystis* sp., *Pseudomonas* sp., *Ralstonia eutropha*, *Rhizobium* sp., and *Rhodobacter sphaeroides* [1]. The substrate for PHAs production was predominantly plant oils (palm and soybean oils are prominent examples), and the utilization of sugars, including edible glucose and inedible lignocellulose is limited [25]. Poly(lactic acid) (PLA) is a commercially available bioplastics, and the starting material of lactic acid is provided by microbial fermentation followed by chemical polymerization [16]. PHAs and PLA are biodegradable polyesters, but the application is limited due to low thermostability (the glass transition temperature, $T_g \leqq 70$ °C) [21].

To overcome the limitation of the existing bioplastics, high-performance of plastics with thermotolerance or high mechanical strength has been synthesized from bio-based chemicals. An aromatic polyester poly(4HCA-*co*-DHCA), which was introduced an plant lignin component of caffeic acid into the main chain, was biodegradable and showed higher thermostability (T_g, 169 °C) compared to PLA [20] (Table 5). The mechanical strength and thermostability of poly(4HCA-*co*-DHCA) were comparable with petroleum-based polycarbonate. A 100% bio-based polyamide, bio-nylon PA5.10, was composed of plant oil derived sebacic acid and fermentation derived diaminopentane, and the glass-fiber showed a comparable mechanical strength with the existing petroleum-based polyamide [31]. Another bio-based polyamide, 10C-6, was derived from the fermentative product of 4-aminocinnamic acid (4ACA), showing a markedly higher mechanical strength and thermostability (σ, 407 MPa; T_g, 243 °C) than those of existing other transparent plastics [59]. An aromatic polyimide, Ami-PBI (80/20), was derived from fermentative products of 3-amino-4-hydroxybenzoic acid (3,4-AHBA) and *p*-aminobenzoic acid (PABA). These two fermentative products were produced from cellulosic biomass, and resulting polymer Ami-PBI (80/20) showed the highest thermostability (T_{10}, 697 °C) than that of all existing plastics [44].

Table 5 Physicochemical properties of bio-based and petroleum-based polymers

Polymer	Type	Biomonomer	Stress, σ (MPa)	Glass transition temp, T_g (°C)	10% weight-loss temp, T_{10} (°C)	References
Aliphatic						
Polyhydroxyalkanoates (PAHs)	Polyester	NA	40	−12	NA	[20]
Poly(lactic acid)	Polyester	Lactic acid	68	55	NA	[20]
Aromatic						
poly(4HCA-*co*-DHCA)	Polyester	Caffeic acid (DHCA) and *p*-coumaric acid (4HCA)	63	169	290	[20]
Bio-nylon PA5.10	Polyamide	Sebacic acid and diaminopentane	NA	50	NA	[31]
10C-6	Polyamide	4-Aminocinnamic acid (4ACA)	407	243	359	[59]
Ami-PBI (80/20)	Polyimide	3-Amino-4-hydroxybenzoic acid (3,4-AHBA) and *p*-aminobenzoic acid (PABA)	63	NA	697	[44]
Petroleum-based						
Polycarbonate	Polyester	NA	62	150	NA	[59]
Nylon 6	Polyamide	NA	75	54	415	[31, 44]
Kapton	Polyimide	NA	231	NA	580	[44]

Abbreviations NA, not applicable

Therefore, the physicochemical properties of bio-based plastics can cover with those of existing petroleum-based plastics.

2.6 Future Perspectives

Biorefinery is integrated technologies composing of plant design, crop cultivation, pretreatment and hydrolyzation of biomass, fermentation, and the separation and purification of products (Fig. 1). This chapter focuses on the production bioplastics from lignocellulosic biomass. Various types of bioplastics with biodegradability or high-performance have been synthesized from biomass (Table 5). The physicochemical properties of thermostability or mechanical strength of the bio-based plastics are comparable with those of commercially available petro-based plastics. However, the share of bioplastics is only 0.6% of the total production of plastics in 2018 (https://www.european-bioplastics.org/market/). To increase the share of bioplastics, further studies are needed for the design and development of biomass plant and improved production of biomonomers from the biomass.

References

1. F. Abd El-malek, H. Khairy, A. Farag, S. Omar, The sustainability of microbial bioplastics, production and applications. Int. J. Biol. Macromol. **157**, 319–328 (2020)
2. M.A. Ali, T. Kaneko, High-performance bioNylons from itaconic and amino acids with pepsin degradability. Adv. Sustain. Syst. (2021)
3. Y. Barrière, C. Riboulet, V. Valérie Méchin, S. Maltese, M. Pichon, A. Cardinal, C. Lapierre, T. Lübberstedt, J.-P. Martinant, Genetics and genomics of lignification in grass cell walls based on maize

as a model system. Genes Genomes Genomics **1**(2), 133–156 (2007)
4. M.F.A. Bradfield, A. Mohagheghi, D. Salvachua, H. Smith, B.A. Black, N. Dowe, G.T. Beckham, W. Nicol, Continuous succinic acid production by *Actinobacillus succinogenes* on xylose-enriched hydrolysate. Biotechnol. Biofuels **8**, 181 (2015)
5. F. Cherubini, N.D. Bird, A. Cowie, G. Jungmeier, B. Schlamadinger, S. Woess-Gallasch, Energy- and greenhouse gas-based LCA of biofuel and bioenergy systems: key issues, ranges and recommendations. Resour. Conserv. Recycl. **53**(8), 434–447 (2009)
6. S.P.S. Chundawat, G.T. Beckham, M.E. Himmel, B.E. Dale, Deconstruction of lignocellulosic biomass to fuels and chemicals. Annu. Rev. Chem. Biomol. Eng. **2**, 121–145 (2011)
7. J. Cotton, G. Burow, V. Acosta-Martinez, J. Moore-Kucera, Biomass and cellulosic ethanol production of forage sorghum under limited water conditions. Bioenergy Res. **6**(2), 711–718 (2013)
8. J.C. da Cruz, A.M. de Castro, E.F.C. Servulo, World market and biotechnological production of itaconic acid. 3 Biotech **8**, 138 (2018)
9. I. Delidovich, P.J.C. Hausoul, L. Deng, R. Pfutzenreuter, M. Rose, R. Palkovits, Alternative monomers based on lignocellulose and their use for polymer production. Chem. Rev. **116**(3), 1540–1599 (2016)
10. Y. Deng, L.Z. Ma, Y. Mao, Biological production of adipic acid from renewable substrates: current and future methods. Biochem. Eng. J. **105**, 16–26 (2016)
11. B.S.Dien, G. Sarath, J.F. Pedersen, S.E. Sattler, H. Chen, D.L. Funnell-Harris, N.N. Nichols, M.A. Cotta, Improved sugar conversion and ethanol yield for forage sorghum (*Sorghum bicolor* L. Moench) lines with reduced lignin contents. Bioenergy Res. **2**(3), 153–164 (2009)
12. K.M. Draths, D.R. Knop, J.W. Frost, Shikimic acid and quinic acid: replacing isolation from plant sources with recombinant microbial biocatalysis. J. Am. Chem. Soc. **121**(7), 1603–1604 (1999)
13. J.R. Elmore, G.N. Dexter, D. Salvachua, J. Martinez-Baird, E.A. Hatmaker, J.D. Huenemann, D.M. Klingeman, G.L. Peabody, D.J. Peterson, C. Singer, G.T. Beckham, A.M. Guss, Production of itaconic acid from alkali pretreated lignin by dynamic two stage bioconversion. Nat. Commun. **12**, 2261 (2021)
14. T. Fujita, H.D. Nguyen, T. Ito, S. Zhou, L. Osada, S. Tateyama, T. Kaneko, N. Takaya, Microbial monomers custom-synthesized to build true bio-derived aromatic polymers. Appl. Microbiol. Biotechnol. **97**(20), 8887–8894 (2013)
15. N.G. Halford, A. Karp, Enegy crops: introduction, in *Enegy Crops*, ed. by N.G. Halford, A. Karp (RSC publishing. Cambridgem UK, 2011), pp. 1–11
16. K. Hamad, M. Kaseem, M. Ayyoob, J. Joo, F. Deri, Polylactic acid blends: the future of green, light and tough. Prog. Polym. Sci. **85**, 83–127 (2018)
17. K.J. Han, W.D. Pitman, M. Kim, D.F. Day, M.W. Alison, M.E. McCormick, G. Aita, Ethanol production potential of sweet sorghum assessed using forage fiber analysis procedures. Global Change Biol. Bioenergy **5**(4), 358–366 (2013)
18. S. Hashimoto, T. Wake, H. Nakamura, M. Minamiyama, S. Araki-Nakamura, K. Ohmae-Shinohara, E. Koketsu, S. Okamura, K. Miura, H. Kawaguchi, S. Kasuga, T. Sazuka, The dominance model for heterosis explains culm length genetics in a hybrid sorghum variety. Sci. Reports **11**(1) (2021)
19. J.L. Hu, Z.T. Zhang, Y.X. Lin, S.M. Zhao, Y.X. Mei, Y.X. Liang, N. Peng, High-titer lactic acid production from NaOH-pretreated corn stover by *Bacillus coagulans* LA204 using fed-batch simultaneous saccharification and fermentation under non-sterile condition. Biores. Technol. **182**, 251–257 (2015)
20. T. Kaneko, T.H. Thi, D.J. Shi, M. Akashi, Environmentally degradable, high-performance thermoplastics from phenolic phytomonomers. Nat. Mater. **5**(12), 966–970 (2006)
21. H. Kawaguchi, T. Hasunuma, C. Ogino, A. Kondo, Bioprocessing of bio-based chemicals produced from lignocellulosic feedstocks. Curr. Opin. Biotechnol. **42**, 30–39 (2016)
22. H. Kawaguchi, T. Hasunuma, Y. Ohnishi, T. Sazuka, A. Kondo, C. Ogino, Enhanced production of γ-amino acid 3-amino-4-hydroxybenzoic acid by recombinant *Corynebacterium glutamicum* under oxygen limitation. Microb. Cell Fact. **20**, 228 (2021)
23. H. Kawaguchi, Y. Katsuyama, D. Danyao, P. Kahar, S. Nakamura-Tsuruta, H. Teramura, K. Wakai, K. Yoshihara, H. Minami, C. Ogino, Y. Ohnishi, A. Kondo, Caffeic acid production by simultaneous saccharification and fermentation of kraft pulp using recombinant *Escherichia coli*. Appl. Microbiol. Biotechnol. **101**(13), 5279–5290 (2017)
24. H. Kawaguchi, K. Sasaki, K. Uematsu, Y. Tsuge, H. Teramura, N. Okai, S. Nakamura-Tsuruta, Y. Katsuyama, Y. Sugai, Y. Ohnishi, K. Hirano, T. Sazuka, C. Ogino, A. Kondo, 3-Amino-4-hydroxybenzoic acid production from sweet sorghum juice by recombinant *Corynebacterium glutamicum*. Biores. Technol. **198**, 410–417 (2015)
25. H. Kawaguchi, K. Takada, T. Elkasaby, R. Pangestu, M. Toyoshima, P. Kahar, C. Ogino, T. Kaneko, A. Kondo, Recent advances in lignocellulosic biomass white biotechnology for bioplastics. Biores. Technol. **344**(Part B), 126165 (2022)
26. H. Kawaguchi, H. Teramura, K. Uematsu, K.Y. Hara, T. Hasunuma, K. Hirano, T. Sazuka, H. Kitano, Y. Tsuge, P. Kahar, S. Niimi-Nakamura, K.I. Oinuma, N. Takaya, S. Kasuga, C. Ogino, A. Kondo, Phenyllactic acid production by simultaneous saccharification and fermentation of pretreated sorghum bagasse. Biores. Technol. **182**, 169–178 (2015)
27. H. Kawaguchi, K. Uematsu, C. Ogino, H. Teramura, S. Niimi-Nakamura, Y. Tsuge, T. Hasunuma, K. Oinuma, N. Takaya, A. Kondo, Simultaneous

saccharification and fermentation of kraft pulp by recombinant *Escherichia coli* for phenyllactic acid production. Biochem. Eng. J. **88**, 188–194 (2014)
28. Y. Kawai, S. Noda, C. Ogino, Y. Takeshima, N. Okai, T. Tanaka, A. Kondo, *p*-Hydroxycinnamic acid production directly from cellulose using endoglucanase- and tyrosine ammonia lyase-expressing *Streptomyces lividans*. Microb. Cell Fact. **12**, 45 (2013)
29. A.A.J. Kerssemakers, P. Domenech, M. Cassano, C.K. Yamakawa, G. Dragone, S.I. Mussatto, Production of itaconic acid from cellulose pulp: Feedstock feasibility and process dtrategies for an efficient microbial performance. Energies **13**(7), 1654 (2020)
30. J.S. Kim, M.N. Islam-Faridi, P.E. Klein, D.M. Stelly, H.J. Price, R.R. Klein, J.E. Mullet, Comprehensive molecular cytogenetic analysis of sorghum genome architecture: distribution of euchromatin, heterochromatin, genes and recombination in comparison to rice. Genetics **171**(4), 1963–1976 (2005)
31. S. Kind, S. Neubauer, J. Becker, M. Yamamoto, M. Volkert, G. von Abendroth, O. Zelder, C. Wittmann, From zero to hero—Production of bio-based nylon from renewable resources using engineered *Corynebacterium glutamicum*. Metab. Eng. **25**, 113–123 (2014)
32. T. Kogure, T. Kubota, M. Suda, K. Hiraga, M. Inui, Metabolic engineering of *Corynebacterium glutamicum* for shikimate overproduction by growth-arrested cell reaction. Metab. Eng. **38**, 204–216 (2016)
33. D. Koma, H. Yamanaka, K. Moriyoshi, K. Sakai, T. Masuda, Y. Sato, K. Toida, T. Ohmoto, Production of *p*-Aminobenzoic acid by metabolically engineered *Escherichia coli*. Biosci. Biotechnol. Biochem. **78**(2), 350–357 (2014)
34. S. Krull, A. Hevekerl, A. Kuenz, U. Pruss, Process development of itaconic acid production by a natural wild type strain of *Aspergillus terreus* to reach industrially relevant final titers. Appl. Microbiol. Biotechnol. **101**(10), 4063–4072 (2017)
35. R.C. Kuhad, A. Singh, Lignocellulose biotechnology: current and future prospects. Crit. Rev. Biotechnol. **13**(2), 151–172 (1993)
36. X. Li, K. Zheng, C.H. Lai, J. Ouyang, Q. Yong, Improved itaconic acid production from undetoxified enzymatic hydrolysate of steam-exploded corn stover using an *Aspergillus terreus* mutant generated by atmospheric and room temperature plasma. BioResources **11**(4), 9047–9058 (2016)
37. X. Li, J. Zhou, S.P. Ouyang, J. Ouyang, Q. Yong, Fumaric acid production from alkali-pretreated corncob by fed-batch simultaneous saccharification and fermentation combined with separated hdrolysis and fermentation at high solids loading. Appl. Biochem. Biotechnol. **181**(2), 573–583 (2017)
38. X.J. Li, Y. Liu, Y. Yang, H. Zhang, H.L. Wang, Y. Wu, M. Zhang, T. Sun, J.S. Cheng, X.F. Wu, L.J. Pan, S.T. Jiang, H.W. Wu, High levels of malic acid production by the bioconversion of corn straw hydrolyte using an isolated *Rhizopus delemar* strain. Biotechnol. Bioprocess Eng. **19**(3), 478–492 (2014)
39. Y.L. Liu, G. Liu, J. Zhang, V. Balan, J. Bao, Itaconic acid fermentation using activated charcoal-treated corn stover hydrolysate and process evaluation based on Aspen plus model. Biomass Convers. Biorefin. **10**(2), 463–470 (2020)
40. C.S. Marvel, T.H. Shepherd, Polymerization reactions of itaconic acid and some of its derivatives. J. Org. Chem. **24**(5), 599–605 (1959)
41. L. Matsakas, P. Christakopoulos, Fermentation of liquefacted hydrothermally pretreated sweet sorghum bagasse to ethanol at high-solids content. Biores. Technol. **127**, 202–208 (2013)
42. N. Mosier, C. Wyman, B. Dale, R. Elander, Y.Y. Lee, M. Holtzapple, M. Ladisch, Features of promising technologies for pretreatment of lignocellulosic biomass. Biores. Technol. **96**(6), 673–686 (2005)
43. M. Murariu, P. Dubois, PLA composites: from production to properties. Adv. Drug Deliv. Rev. **107**, 17–46 (2016)
44. A. Nag, M.A. Ali, H. Kawaguchi, S. Saito, Y. Kawasaki, S. Miyazaki, H. Kawamoto, D.T.N. Adi, K. Yoshihara, S. Masuo, Y. Katsuyama, A. Kondo, C. Ogino, N. Takaya, T. Kaneko, Y. Ohnishi, Ultrahigh thermoresistant lightweight bioplastics developed from fermentation products of cellulosic feedstock. Adv. Sustain. Syst. **5**, 2000193 (2021)
45. A. Nag, M.A. Ali, M. Watanabe, M. Singh, K. Amornwachirabodee, S. Kato, T. Mitsumata, K. Takada, T. Kaneko, High-performance poly(benzoxazole/benzimidazole) bio-based plastics with ultra-low dielectric constant from 3-amino-4-hydroxybenzoic acid. Polym. Degrad. Stab. **162**, 29–35 (2019)
46. R.L. Ordonio, Y. Ito, A. Hatakeyama, K. Ohmae-Shinohara, S. Kasuga, T. Tokunaga, H. Mizuno, H. Kitano, M. Matsuoka, T. Sazuka, Gibberellin deficiency pleiotropically induces culm bending in sorghum: an insight into sorghum semi-dwarf breeding. Sci. Rep. **4**, 5287 (2014)
47. L.L. Peng, N.Z. Xie, L. Guo, L.M. Wang, B. Yu, Y.H. Ma, Efficient open fermentative production of polymer-grade L-lactate from sugarcane bagasse hydrolysate by thermotolerant *Bacillus* sp strain P38. PLoS ONE **9**(9), e107143 (2014)
48. L.A. Rivera-Burgos, J.J. Volenec, G. Ejeta, Biomass and bioenergy potential of brown midrib sweet sorghum germplasm. Front. Plant Sci. **10**, 1142 (2019)
49. T. Robert, S. Friebel, Itaconic acid: a versatile building block for renewable polyesters with enhanced functionality. Green Chem. **18**(10), 2922–2934 (2016)
50. W.L. Rooney, J. Blumenthal, B. Bean, J.E. Mullet, Designing sorghum as a dedicated bioenergy feedstock. Biofuels, Bioprod. Biorefin. **1**(2), 147–157 (2007)
51. A. Saballos, G. Ejeta, E. Sanchez, C. Kang, W. Vermerris, A Genomewide Analysis of the cinnamyl

alcohol dehydrogenase family in Sorghum [*Sorghum bicolor* (L.) Moench] identifies *SbCAD2* as the *brown midrib6* gene. Genetics **181**(2), 783–795 (2009)
52. K. Sasaki, Y. Tsuge, D. Sasaki, H. Kawaguchi, T. Sazuka, C. Ogino, A. Kondo, Repeated ethanol production from sweet sorghum juice concentrated by membrane separation. Biores. Technol. **186**, 351–355 (2015)
53. A. Sawisit, S.S. Jantama, S. Kanchanatawee, K. Jantama, Efficient utilization of cassava pulp for succinate production by metabolically engineered *Escherichia coli* KJ122. Bioprocess Biosyst. Eng. **38**(1), 175–187 (2015)
54. I. Schlembach, H.H. Tehrani, L.M. Blank, J. Büchs, N. Wierckx, L. Regestein, M.A. Rosenbaum, Consolidated bioprocessing of cellulose to itaconic acid by a co-culture of Trichoderma reesei and *Ustilago maydis*. Biotechnol. Biofuels **13**, 207 (2020)
55. B. Sriputorn, P. Laopaiboon, L. Laopaiboon, Increase in ethanol production efficiency from sweet sorghum juice by *Saccharomyces cerevisiae*. Chiang Mai J. Sci. **44**(1), 59–71 (2017)
56. H. Suzuki, Y. Ohnishi, Y. Furusho, S. Sakuda, S. Horinouchi, Novel benzene ring biosynthesis from C_3 and C_4 primary metabolites by two enzymes. J. Biol. Chem. **281**(48), 36944–36951 (2006)
57. Y. Tachibana, T. Masuda, M. Funabashi, M. Kunioka, Chemical synthesis of fully biomass-based poly(butylene succinate) from inedible-biomass-based furfural and evaluation of itsbiomass carbon ratio. Biomacromolecules **11**(10), 2760–2765 (2010)
58. T. Tateno, Y. Okada, T. Tsuchidate, T. Tanaka, H. Fukuda, A. Kondo, Direct production of cadaverine from soluble starch using *Corynebacterium glutamicum* coexpressing alpha-amylase and lysine decarboxylase. Appl. Microbiol. Biotechnol. **82**(1), 115–121 (2009)
59. S. Tateyama, S. Masuo, P. Suvannasara, Y. Oka, A. Miyazato, K. Yasaki, T. Teerawatananond, N. Muangsin, S.M. Zhou, Y. Kawasaki, L.B. Zhu, Z.M. Zhou, N. Takaya, T. Kaneko, Ultrastrong, transparent polytruxillamides derived from microbial photodimers. Macromolecules **49**(9), 3336–3342 (2016)
60. B.-E. Teleky, D.C. Vodnar, Biomass-derived production of itaconic acid as a building block in specialty polymers. Polymers **11**(6) (2019)
61. N. Tippkötter, A.M. Duwe, S. Wiesen, T. Sieker, R. Ulber, Enzymatic hydrolysis of beech wood lignocellulose at high solid contents and its utilization as substrate for the production of biobutanol and dicarboxylic acids. Biores. Technol. **167**, 447–455 (2014)
62. J. Tomaszewska, D. Bieliński, M. Binczarski, J. Berlowska, P. Dziugan, J. Piotrowski, A. Stanishevskye, I.A. Witońska, Products of sugar beet processing as raw materials for chemicals and biodegradable polymers. RCS Adv. **8**, 3161–3177 (2018)
63. D.R. Vardon, M.A. Franden, C.W. Johnson, E.M. Karp, M.T. Guarnieri, J.G. Linger, M.J. Salm, T.J. Strathmann, G.T. Beckham, Adipic acid production from lignin. Energy Environ. Sci. **8**(2), 617–628 (2015)
64. J. Wang, B. Roe, S. Macmil, Q. Yu, J.E. Murray, H. Tang, C. Chen, F. Najar, G. Wiley, J. Bowers, M.-A. Van Sluys, D.S. Rokhsar, M.E. Hudson, S.P. Moose, A.H. Paterson, R. Ming, Microcollinearity between autopolyploid sugarcane and diploid sorghum genomes. BMC Genomics **11**, 261 (2010)
65. L.J. Wang, Z.L. Luo, A. Shahbazi, Optimization of simultaneous saccharification and fermentation for the production of ethanol from sweet sorghum (Sorghum bicolor) bagasse using response surface methodology. Ind. Crops Prod. **42**, 280–291 (2013)
66. Y. Wang, C.J. Chen, D. Cai, Z. Wang, P.Y. Qin, T.W. Tan, The optimization of L-lactic acid production from sweet sorghum juice by mixed fermentation of *Bacillus coagulans* and *Lactobacillus rhamnosus* under unsterile conditions. Biores. Technol. **218**, 1098–1105 (2016)
67. M. Watanabe, M. Makino, N. Kaku, M. Koyama, K. Nakamura, K. Sasano, Fermentative L-(+)-lactic acid production from non-sterilized rice washing drainage containing rice bran by a newly isolated lactic acid bacteria without any additions of nutrients. J. Biosci. Bioeng. **115**(4), 449–452 (2013)
68. T. Werpy, P. Petersen, Top Value Added Chemicals Biomass Volume I: Results of Screening for Potential Candidates from Sugars and Synthesis Gas, U.S. Department of Energy. Oak Ridge, TN (2004)
69. X.F. Wu, Q. Liu, Y.D. Deng, J.H. Li, X.J. Chen, Y.Z. Gu, X.J. Lv, Z. Zheng, S.T. Jiang, X.J. Li, Production of itaconic acid by biotransformation of wheat bran hydrolysate with *Aspergillus terreus* CICC40205 mutant. Biores. Technol. **241**, 25–34 (2017)
70. J. Yang, H. Xu, J.C. Jiang, N. Zhang, J.C. Xie, J. Zhao, Q. Bu, M. We, Itaconic acid production from undetoxified enzymatic hydrolysate of bamboo residues using *Aspergillus terreus*. Biores. Technol. **307**, 123208 (2020)
71. J.L. Yu, T. Zhang, J. Zhong, X. Zhang, T.W. Tan, Biorefinery of sweet sorghum stem. Biotechnol. Adv. **30**(4), 811–816 (2012)
72. J.Z. Zhang, D. Cai, Y.L. Qin, D.H. Liu, X.B. Zhao, High value-added monomer chemicals and functional bio-based materials derived from polymeric components of lignocellulose by organosolv fractionation. Biofuels Bioproducts Biorefining-Biofpr **14**(2), 371–401 (2020)
73. C. Zhao, S.L. Chen, H. Fang, Consolidated bioprocessing of lignocellulosic biomass to itaconic acid by metabolically engineering *Neurospora crassa*. Appl. Microbiol. Biotechnol. **102**(22), 9577–9584 (2018)
74. Z.B. Zheng, T. Chen, M.N. Zhao, Z.W. Wang, X.M. Zhao, Engineering *Escherichia coli* for succinate production from hemicellulose via consolidated bioprocessing. Microb. Cell Fact. **11**, 37 (2012)
75. J. Zhou, J. Ouyang, M. Zhang, H. Yu, Simultaneous saccharification and fermentation of bagasse sulfite pulp to lactic acid by *Bacillus coagulans* CC17. BioResources **9**(2), 2609–2620 (2014)

Syntheses of Biobased Polymers Using Bio/Naturally Derived Products

Kenji Takada and Tatsuo Kaneko

1 Production of Biobased Polymers

In recent years, due to concerns about global warming and the depletion of petroleum resources, the need for a "low-carbon society" that suppresses the emission of carbon dioxide, which accounts for a large proportion of greenhouse gases, has been emphasized [1–5]. And, as one of the solutions to realize a low-carbon society, the material conversion to change the plastic usage rate of excessive industrial products has been proposed. However, the reason why plastics have become so widespread in the world is that they are extremely excellent in terms of both functionality and productivity. This tendency is particularly remarkable in general-purpose plastics such as polyethylenes, poly(vinyl chloride)s, polypropylenes, polystyrenes, and polymethacrylates. However, even if we aim to replace biomass plastics for these de-petroleum and low carbonization, it is not realistic to replace them with petroleum because the bioengineering production process is costly in the current situation. As an economic method, there is a movement to incorporate low carbon as one of the commercial values along with quality, performance, price, etc., and to reflect the fact that it is low carbon in the price to support its spread, but it is still fundamental. The disadvantage of high cost is that it is a heavy shackle, and it is difficult for non-environmentally conscious consumers to penetrate it. While polyesters such as polylactides, poly(ε-caprolactone)s, and poly(butylene succinate)s are attracting attention as plastics that can be decomposed even in the environment, problems related to durability are often pointed out. At present, the use of these typical polyesters is limited due to the problem of durability. The reason for this is that the structures constituting the polymer do not include structures such as aromatic rings, heteroatoms, reactive functional groups, and π-conjugated structures that impart high strength to the materials. Therefore, there is a demand for the development of useful new high-performance and which were obtained from functional biomass.

K. Takada · T. Kaneko
Graduate School of Advanced Science and Technology, Japan Advanced Institute of Science and Technology, 1-1 Asahidai, Nomi 923-1292, Ishikawa, Japan
e-mail: takada@jaist.ac.jp

T. Kaneko (✉)
School of Chemical and Material Engineering, Jiangnan University, Wuxi 214122, Jiangsu, China
e-mail: tkaneko@jiangnan.edu.cn

T. Kaneko (ed.), *Photo-switched Biodegradation of Bioplastics in Marine Environments*,
https://doi.org/10.1007/978-981-99-4354-8_3

2 Polycinnamate as Biobased Aromatic Polyester

Coumaric acids such as 4-hydroxycinnamic acid, 3-methoxy-4-hydroxycinnamic acid (ferulic acid), and 3,4-dihydroxycinnamic acid (caffeic acid) are a type of phenolic plant compounds and contain photosensitive phenyl groups and hydroxy acids in their structure (Fig. 1). These are a polyphenol-based aromatic polyfunctional substance that constitutes the cell wall of plants and the like. Plants consist of cellulose, hemicellulose and lignin, and in particular, lignin is a bio-derived aromatic substance. However, the molecular structure of lignin is still unclear, and its use as a material.

On the other hand, many cinnamic acids, which are biosynthetic precursors of the lignin, naturally have aromatic rings and π-conjugated structures in their molecular structures and have a large number of functional groups on the aromatic rings. It is also easy to establish a production system using enzymes and extraction from plants. Since cinnamic acid has a hydroxy group and a carboxyl group in its structure, it is polymerized by a polycondensation reaction. These molecules are present in the biosynthetic pathway of plant lignin and are known as photo-active yellow proteins as constituent proteins of some photosynthetic bacteria [10, 11]. Furthermore, since the enzymatic synthesis pathway of phytomonomers using amino acids as a starting material has already been established in the field of biochemistry, mass production is possible by a simple method.

2.1 High-Performance Polymers

Polyester synthesized with 4-hydroxycinnamic acid has been found to exhibit high mechanical strength and liquid crystallinity due to its rigid molecular main chain [12–16]. However, due to its rigid molecular structure, it is difficult to increase the molecular weight, and any poly(4-hydroxycinnamic acid) (P4HCA) exhibits brittle properties. Copolymer of P4HCA and poly(3,4-dihydroxycinnamic acid) (PdHCA), P(4HCA-*co*-dHCA), whose rigidity is softened and its rigidity was improved by copolymerizing this polymerization system using cinnamic acids having a branched structure was obtained (Fig. 2). Since these poly(hydroxycinnamic acid)s have very high crystallinity, they are biodegradable but very low efficiency in degradability. However, it was confirmed that by copolymerizing with PdHCA, these crystallinities become lowered, and the biodegradability is partially improved. Furthermore, by utilizing the property of the cinnamic acid unit

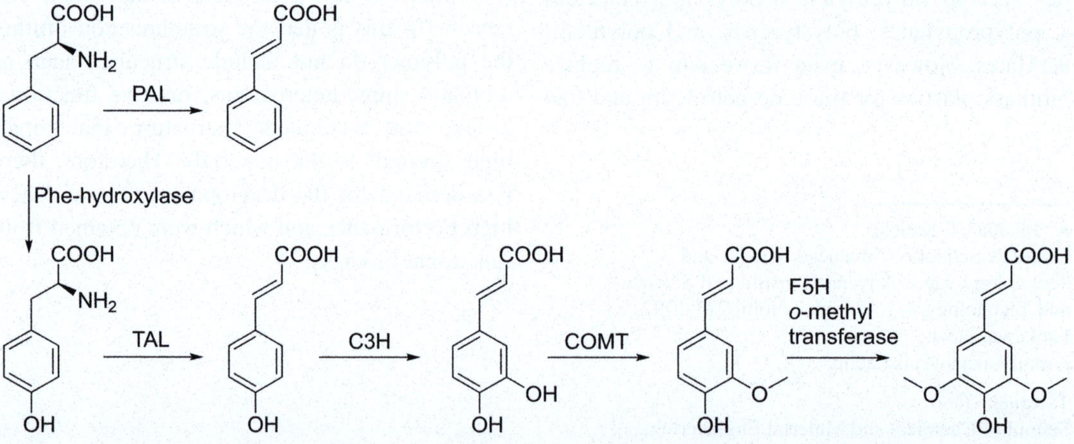

Fig. 1 Biologically synthetic pathway of various cinnamic acids: PAL, phenylalanine ammonia-lyase [6]; TAL, tyrosine ammonia-lyase [6]; C3H, *p*-coumarate 3-hydroxylase [7]; COMT, 2-hydroxypropyl-CoM lyase [8]; F5H, ferulate 5-hydroxylase [9]

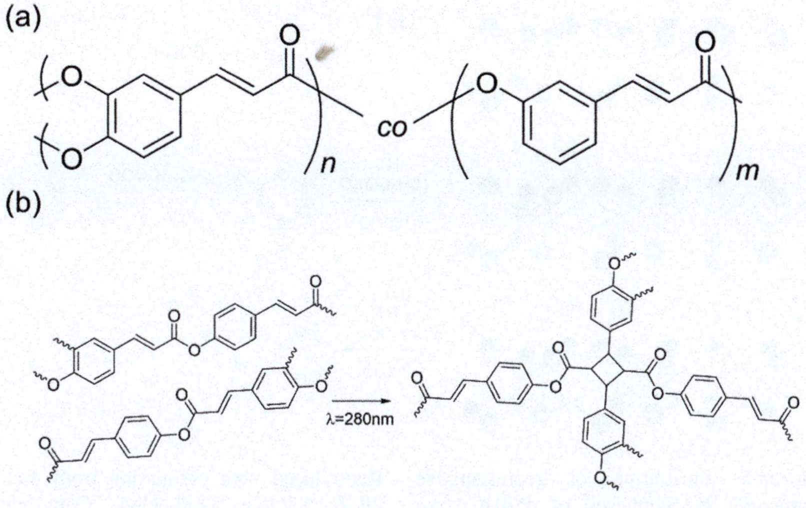

Fig. 2 **a** Structure of biobased polyester P(4HCA-*co*-dHCA). **b** Photocrosslinking of P(4HCA-*co*-dHCA) by UV rays [17]

to cause a [2+2] cycloaddition reaction by ultraviolet (UV) rays, it forms the cyclobutane from the internal olefin of the main chain and greatly reduces the crystallinity. As a result, P(4HCA-*co*-dHCA) after UV irradiation shows very fast soil degradability [17].

A base is used as the catalyst for the synthesis of these poly(hydroxycinnamic acid)s. In the case of hydrotalcite (HT), which is a kind of natural mineral, gives a higher molecular weight poly(hydroxycinnamic acid)s. Copolymer of P(4HCA-*co*-dHCA) prepared in the presence of HT catalyst showed high molecular weight and a narrow molecular weight distribution. A concentration of 0.6 wt% was a suitable ratio for the preparation of the copolymer with high yields and molecular weight. In addition, when HT with an Mg/Al ratio of 3/1 was used, the value of the copolymer of weight-average molecular weight (M_w) showed the highest value. Since HT with an Mg/Al ratio of 3/1 had the highest structural order in the layer structure, the layer structure may be closely correlated with the catalytic activity of acid degradation. P(4HCA-*co*-dHCA) prepared in the presence of HT showed some clear threads. These results indicate that the presence of an HT in P(4HCA-*co*-dHCA) formed a clear interface in the orientation domain (Fig. 3).

This is caused by the fineness of the copolymer structure. Cross-polarization microscopy reveals that the polymer chains are oriented to the surface of the glass fibers and that the resin mixed with the glass fiber fillers aligned along the vertical axis showed very high mechanical strength. In addition, the hybrid resin annealed at 300 °C for 20 min exhibited a softening temperature of 305 °C while maintaining a high mechanical strength of 85 MPa and a high Young's modulus of over 1 GPa (Fig. 4). These values were high enough to be used as biobased high-performance plastics [18].

2.2 Biobased Adhesive

In addition, dHCA has a catechol group in its structure, which is known to be an important functional group for developing strong adhesion of 3-(3,4-dihydroxyphenyl)-**L**-alanine (DOPA), which is an adhesive protein of mussel (Fig. 5). This dHCA-based hyperbranched polyester also exhibits high adhesive performance due to the catechol group at each chain end. However, since only use of dHCA showed a low degree of polymerization, when 4-hydroxycinnamic acid is used as the comonomer, a polyester having a

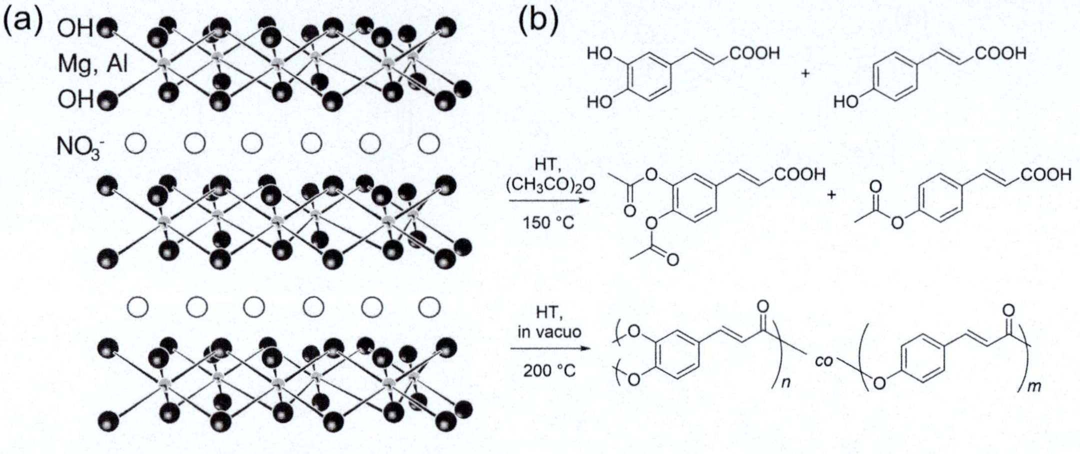

Fig. 3 **a** Schematic illustration of representative hydrotalcite structure. **b** Syntheses of P(4HCA-*co*-dHCA) from acetylated monomers via acidolysis. Reproduced with permission from *Adv. Funct. Mater.*, **2012**, *22(16)*, 3438–3444. Copyright 2012 Wiley Publishing Group [18]

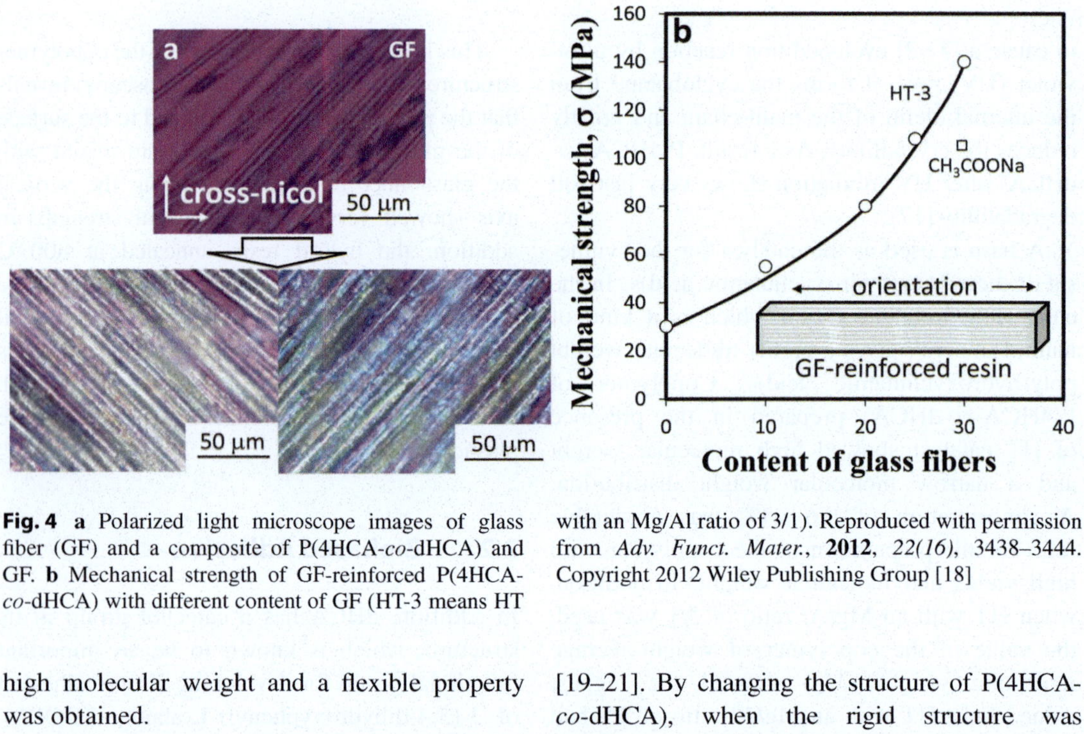

Fig. 4 **a** Polarized light microscope images of glass fiber (GF) and a composite of P(4HCA-*co*-dHCA) and GF. **b** Mechanical strength of GF-reinforced P(4HCA-*co*-dHCA) with different content of GF (HT-3 means HT with an Mg/Al ratio of 3/1). Reproduced with permission from *Adv. Funct. Mater.*, **2012**, *22(16)*, 3438–3444. Copyright 2012 Wiley Publishing Group [18]

high molecular weight and a flexible property was obtained.

The polyester thus obtained showed strong adhesion to the surface of organic and inorganic materials. The adhesive strength of the resin was evaluated by lap shear adhesion testing. The origin of this strong adhesive property is due to the strong hydrogen bond interaction between these polyesters and the oxidation and/or OH groups on each substrate [19–21]. By changing the structure of P(4HCA-*co*-dHCA), when the rigid structure was decreased the adhesive property was increased. Poly((3,4-dihydroxyphenylpropionic acid)-*co*-(4-hydroxyphenylpropionic acid) (P(dHPA-*co*-4HPA)) showed strong adhesive properties for various materials, and its adhesive strength was highly dependent on the dHPA composition. These adhesive strengths are higher than the corresponding values of commercial instant

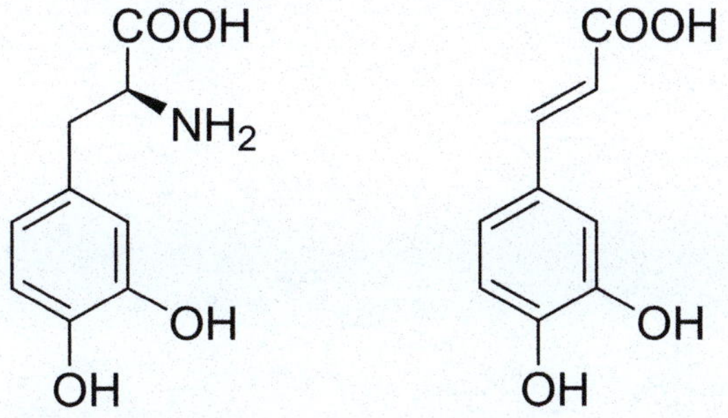

Fig. 5 Chemical structure of DOPA (left) and dHCA (right)

adhesives. In addition, the structural effects of the 4HPA and 3-hydroxyphenylpropionic acid (3HPA) moieties on the properties of the prepared polymers were studied. P(dHPA-*co*-4HPA) containing the *para*-substituted benzene moiety showed a clearly improved T_g compared to P(dHPA-*co*-3HPA) containing meta-substituted benzene moiety at the same composition. The difference in adhesive strength between P(dHPA-*co*-4HPA) and P(dHPA-*co*-3HPA) is probably due to the high symmetry and low flexibility of P(dHPA-*co*-4HPA). Featuring rigid parts, strong adhesiveness, and properly improved T_gs, P(dHPA-*co*-4HPA) is expected to have potential applications in a variety of medical fields, including dental adhesives [22].

2.3 Photoresponsive Materials

Light such as UV rays and visible light is useful for its remoteness and controllability (wavelength, intensity, direction), and is one of the familiar energies. Research on linking such light energy to chemistry and materials is wide-ranging, and photoluminescence using light energy itself, use of heat energy focusing on the effect of light heating, use as electrical energy mediated by solar cells, etc., and materials that are deformed by light are examples, and there are various uses. On the other hand, the development of actuators that change shape and size in response to external stimuli is a theme that has attracted the attention of many researchers since the concept was conceived [23–25]. Actuators are well-known as energy transducers that can convert energy input in various forms into physical motion. In addition, materials for actuators have been actively developed so that various external stimuli can be received efficiently. The abovementioned "light" as one of the external stimuli for these actuators is also attracting attention. When a substance absorbs light and changes its shape or volume, it will convert light energy directly into mechanical work, which is nothing but a single-step direct energy conversion. If simple mechanical energy direct conversion can be realized using such materials and light, useful effects such as system miniaturization and reduction of energy loss are to be expected. In addition, it can be expected to be easy to handle materials due to its characteristics such as operability and quick response by the optical remote used for external stimuli, and it is expected to be widely applied in fields such as industry and medical treatment. These materials are achieved by introducing light-sensitive functional groups into the polymer structure, and various functional group and introduction methods have been studied. There are many reports of material development aimed at photodeformation, and many reports have been given on gels [26] and elastomers [27] even for polymer-based materials. In recent years, it has been reported that the isomerization reaction of photochromic molecules is combined with the cooperative phenomenon of liquid crystals to realize macroscopic molecular orientation as a liquid crystal elastomer (Fig. 6) [28–31].

Under such a situation, there are only reports on the fact that the previously reported photodeformable materials are only soft materials and the photoreaction of polymer side chains of cinnamate moiety. The development of soft actuators, which is one of the goals of soft materials, has the great advantage of being light and soft and having high elasticity, unlike hard actuators using ceramics and metals. In addition, when we expect it to be applied to artificial muscles and medical materials as a destination, the natural biological actuators that show functions in our body are organized by hydrogel composed of 90% or more water. Therefore, it

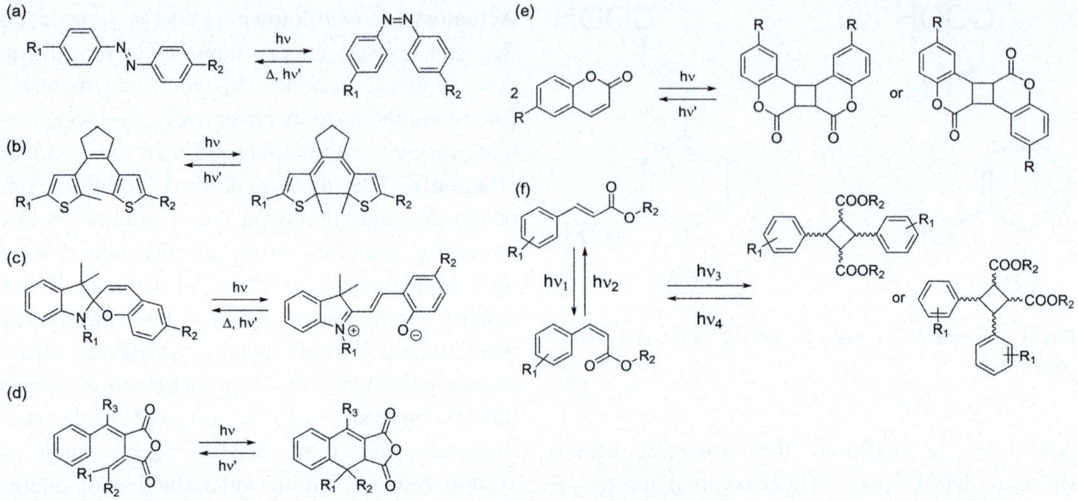

Fig. 6 Candidate of the functional group introduced into polymers with photoresponsiveness: **a** azobenzene, **b** dithienylethene (diarylethene), **c** spiropyran, **d** flugide, **e** coumarin, **f** cinnamic acid [28–31]

is inevitable that softer materials will become the mainstream of photostimulus-responsive materials.

Cinnamic acids are also attracting attention as a photoresponsive functional group. Focusing on the structure of this main chain type cinnamic acid-based polyester, since it has a cinnamoyl group as a photoreactive group, it has two reactions: the *E-Z* (*trans-cis*) isomerization of the intramolecular reaction and the [2+2] cycloaddition of the intermolecular reaction [32–34]. It turns out that it is a photochromic molecule that exhibits a kind of photoreactivity. To give an example in which such a photoresponse based on the cinnamic acid structure is used in nature, there is a case where a photosensing protein of a photosynthetic microorganism utilizes the isomerization of hydroxycinnamic acid for the tertiary structure change of the protein [10, 11].

The acidolysis polymerization method by acetylating and deacetylating of hydroxyl groups using 4HCA, 3HCA, 3-methoxy-4-hydroxycinnamic acid (3M4HCA), and dHCA as various hydroxycinnamic acids to afford the corresponding polyesters (Fig. 7) [15–17, 35, 36]. These methods of polymer synthesis were often performed in a solvent-free system for the purpose of improving environmental friendliness and reactivity, and this method also employed as synthesizing high-performance polyester "Vectra". In this polymerization method, the monomer is generally polymerized by heating and stirring the monomer at about 200 °C under vacuum conditions in the presence of sodium acetate and acetic anhydride as a catalyst and an acetylation agent, respectively. It was confirmed that in the case of hydroxycinnamic acid, it is possible to synthesize a polyester in the same manner as these.

These poly(cinnamic acid)s also be molded, although under suitable conditions. For instance, P3HCA and PdHCA are soluble in amide-based and strong acid solvents such as *N,N*-dimethylformamide (DMF), dimethyl sulfoxide (DMSO) and trifluoroacetic acid (TFA) [38]. Therefore, a hard film of P3HCA could be easily obtained by the solution casting method. When this cast film was cut into strips (20 × 8 × 0.1 mm) and irradiated with UV, it was confirmed that the film was bent (Fig. 8).

Irradiation of these polymers with UV (280–450 nm) confirmed a decrease in absorbance with time to irradiation. As a result of infrared (IR) measurement of the surface of the film before and after photodeformation of P3HCA and PdHCA, which could be formed into a ductile film, a peak was found on the

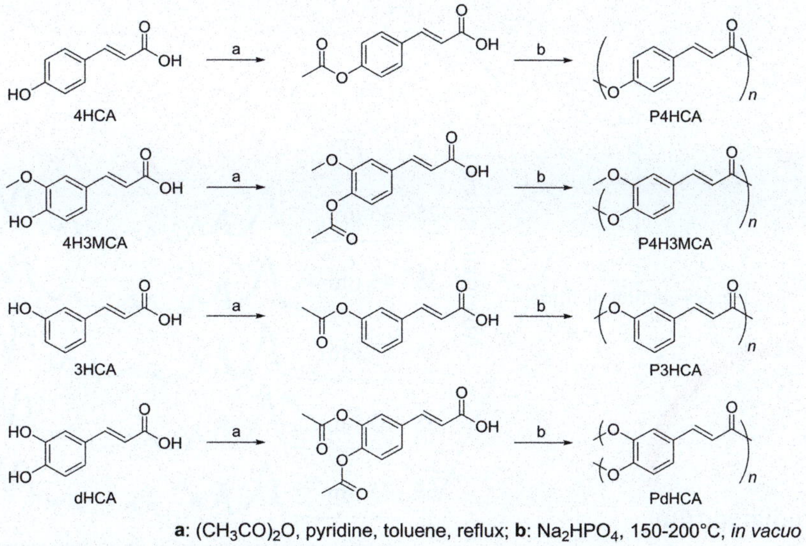

Fig. 7 Synthesis of polycinnamates by acetylation and acidolysis polycondensation [37]

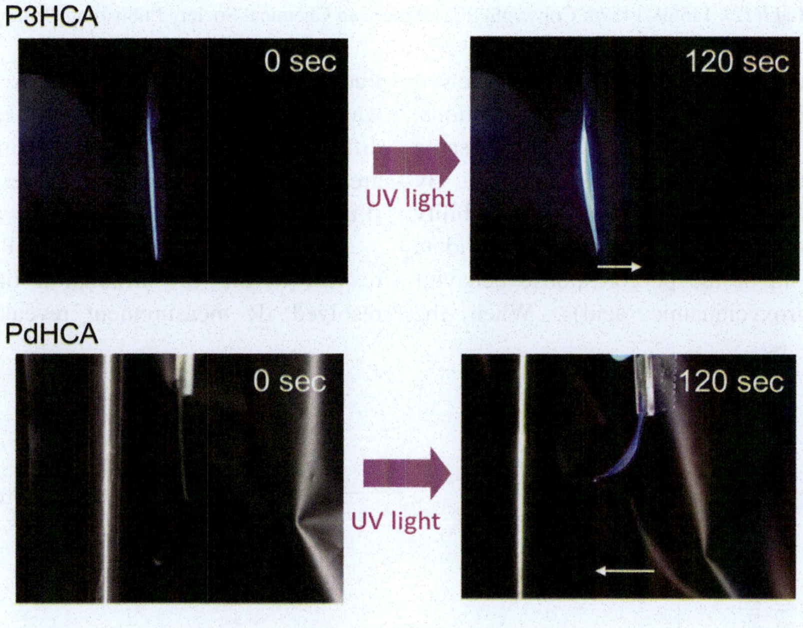

Fig. 8 Deformation behavior of P3HCA and PdHCA film with UV (high-pressure mercury lamp, 280–450 nm). The deformation direction is indicated by the arrow direction [37]

UV irradiation surface. A decrease in peak intensity derived from vinyl groups of 1640 and 980 cm$-^{1}$ was observed. Furthermore, in order to confirm the optical change of the internal structure of the film in more detail, the photo-deformed P3HCA film was cut with a diamond cutter, and IR mapping was attempted at 0.01 mm intervals on the cut surface (Fig. 9).

As a result, it was confirmed that UV irradiation affects to 20 μm depth from the surface of the film. It caused the change in the structure of these cinnamate-based polyesters.

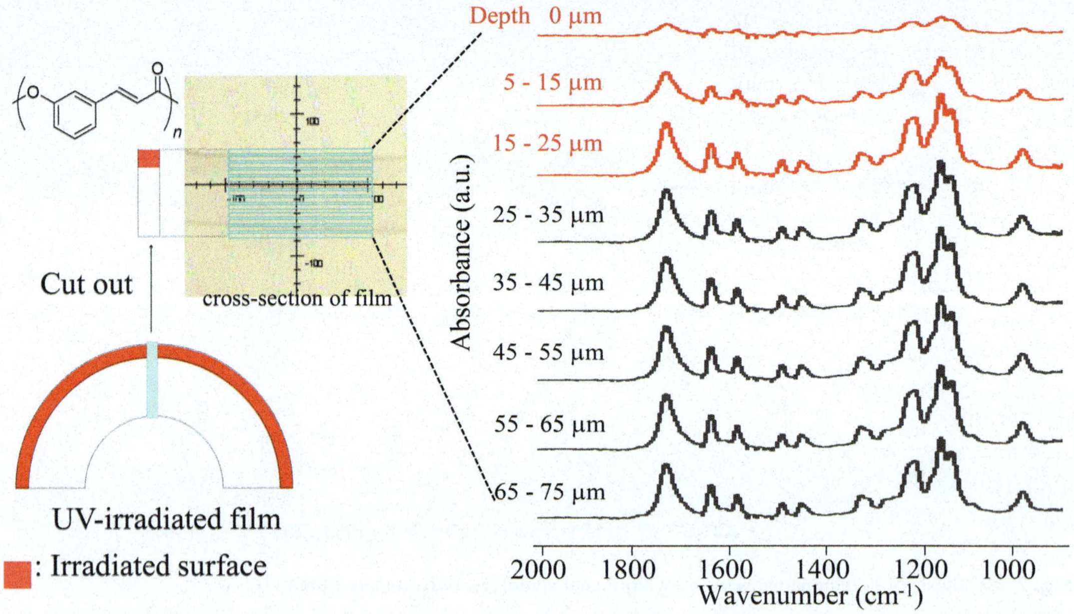

Fig. 9 IR spectra of a cross-section of P3HCA cut film. Reproduced with permission from *ACS Appl. Mater. Interfaces* **2021**, *13(12)*, 14569–14576. Copyright 2021 American Chemical Society Publishing Group [37]

Furthermore, P3HCA and PdHCA showed convex and concave film deformations, respectively, even though they had the same cinnamic acid unit in the polymer unit. By elucidating these differences in deformability by step-scan analysis, it is possible to elucidate the mechanism of the photoresponse behavior of poly(hydroxycinnamic acid)s. When the fluorescence lifetime of P3HCA and PdHCA was measured, the types of excited states differed between them, suggesting that there are two types of excited states in P3HCA (Fig. 10).

A structural analysis of P3HCA with respect to the UV irradiation time by time-resolved IR measurement revealed that only

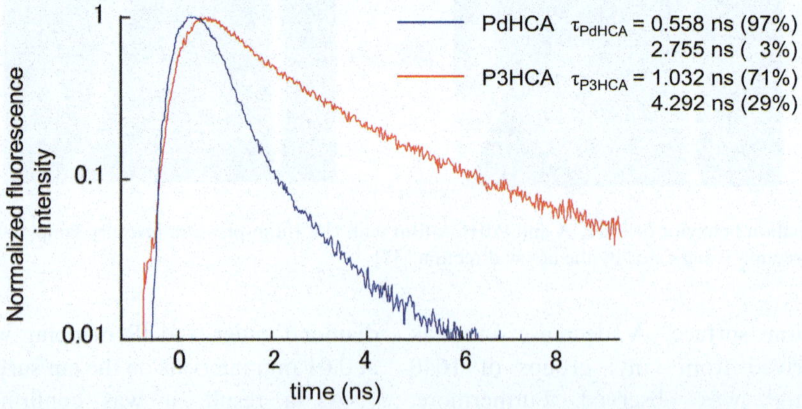

Fig. 10 Time-resolved fluorescence lifetime decay profiles of the PdHCA and P3HCA films in the bulk state. Reproduced with permission from *ACS Appl. Mater. Interfaces* **2021**, *13(12)*, 14569–14576. Copyright 2021 American Chemical Society Publishing Group [37]

P3HCA causes *cis* isomerization by UV irradiation, and the reason why the P3HCA film shows convex deformation is that the P3HCA film shows convex deformation in the polymer chains (Fig. 11). It was found that due to "photoexpansion" associated with *cis* isomerization [37]. The optical deformation of the P3HCA film that shows this convex deformation lifts it up even when a glass rod is placed as a load, and it is a powerful one with unprecedented micro-Newton order work [39, 40]. From this result, it can be concluded that the optical deformation of the film is the result of the optical change of the molecular structure generated hierarchically inside the film. Regarding the fact that the penetration of UV into the film and partial photodeformation affect the bending of the entire film, similar reports have been found in the reports on photodeformation in liquid crystal elastomers [41, 42], and the outside of light. It is considered to be appropriate as a deformation mechanism of the film in response to the stimulus. In addition, since the thickness of the optical structure change is as thin as 20 μm, it may be possible to induce optical deformation by coating the surface of other flexible and versatile substances. These polyesters are expected future prospects such as the ability to make any plastic film optical functional.

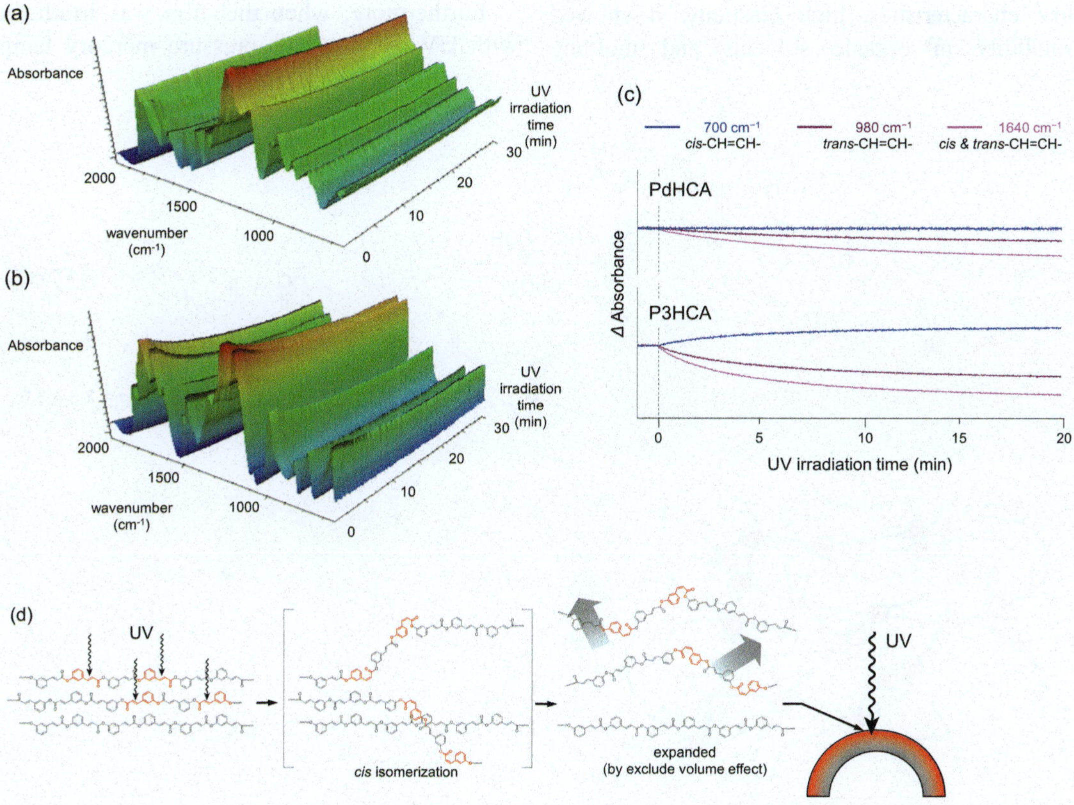

Fig. 11 3D image of step-scanning IR spectra during UV irradiation of **a** PdHCA and **b** P3HCA. **c** Time-resolved differential of the IR spectra of the PdHCA and P3HCA films based on the step-scanning IR spectra picked up the absorbance of 700, 980, 1640 cm^{-1}. **d** Schematic illustration of the structural changes in P3HCA; photoexpansion by UV irradiation. Reproduced with permission from *ACS Appl. Mater. Interfaces* **2021**, *13(12)*, 14569–14576. Copyright 2021 American Chemical Society Publishing Group [37]

2.4 Shape Memory Materials

Not only the homopolymerization of hydroxycinnamic acid, but also the copolymerization of poly(cinnamic acid)s by introducing a rigid structure into the main chain and aliphatic diacid and glycol as a flexible chain is excellent for obtaining an optical shape memory material [43]. For instance, by reacting methyl 4-hydroxycinnamic acid with succinic acid to synthesize a dicarboxylic acid monomer having two cinnamic acid moieties, and polymerizing this with various glycols with different chain numbers in the presence of a germanium dioxide as a catalyst to afford a highly branched polyester (P4HCA-Su) with a degree of branching of or more was obtained (Fig. 12) [43]. While the obtained highly branched polyester showed rubber-like characteristics, high elasticity, it showed solubility in organic solvents and molding processability not found in ordinary rubbers. It is considered that the reason for obtaining such unique properties is that this polymer was obtained as a highly branched polymer entangled by the bulk transesterification method. It is generally said that the viscosity of a highly branched polymer is remarkably lowered because entanglement between polymer chains hardly occurs, and its molding process is difficult. In addition, the network polymers exhibit a shape memory effect like rubber, but it is difficult to perform molding such as hot pressing because the molecular chain does not exhibit fluidity. However, the obtained cinnamate-based polyester has become a branched structure that was molded by hot pressing and has a shape memory effect because a highly branched structure is formed while the molecular chains are entangled to form an elastomer having molding processability.

Furthermore, when this film was irradiated with UV using a high-pressure mercury lamp,

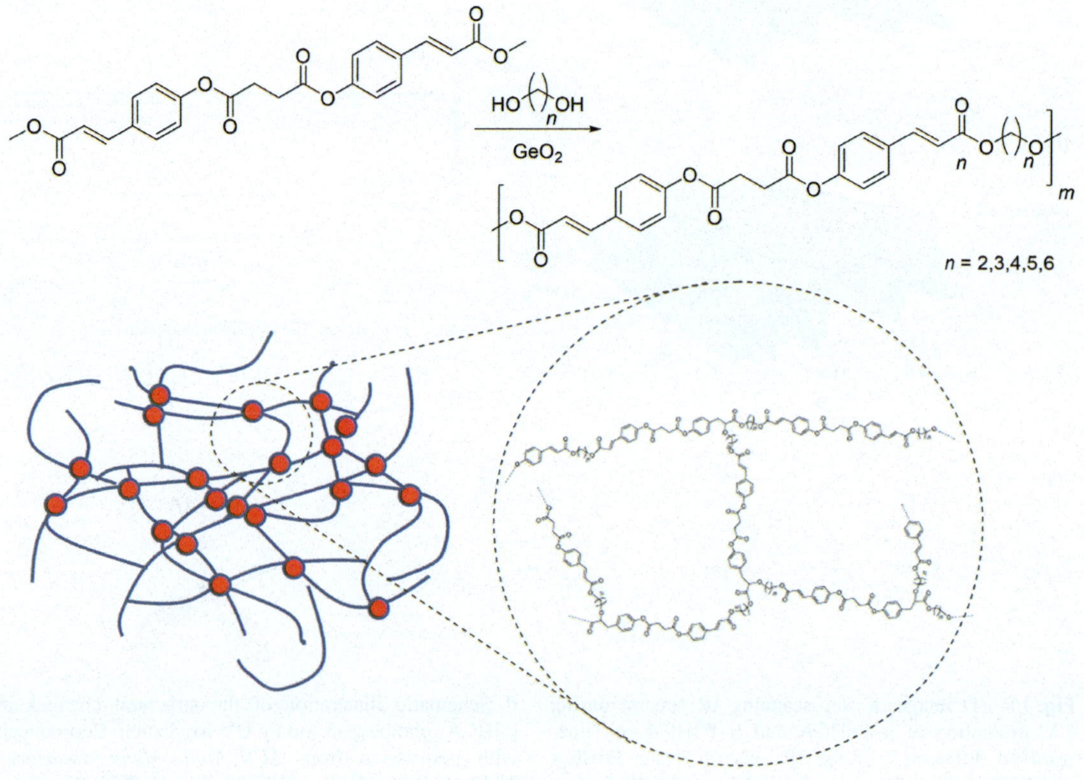

Fig. 12 Synthesis of copolymers of 4HCA, succinic acid, and various aliphatic diols [43]

it showed a mechanistic deformation effect of bending with the irradiated side as a concave surface. In this case, it was confirmed that the deformation speed depends on the carbon number of the glycol to be introduced, and that the deformation speed becomes slower when the carbon number is small, and the flexible chain site is short. It was also found that the shape after optical deformation is memorized, and that this shape is used as the initial shape to show the shape restoration effect by temperature. Furthermore, it was found that irradiation with UV-C as a shorter wavelength (<250 nm) from a low-pressure mercury lamp restores the original flat film (Fig. 13). Furthermore, this property expected the possibility of using it as a multi-shape memory film that can store various shapes with multiple stimuli.

3 Biobased High-Performance Polymers Using Amino Acid as Diamine

3.1 Hetero Diamine-Based Polymers

The development of high-performance plastics using naturally and biologically produced raw materials as starting materials is an effective means for realizing a sustainable society, and carbon dioxide-derived carbon in the atmosphere is incorporated into polymer materials for a long period of time. This approach is expected that fixed and the carbon dioxide concentration in the atmosphere be continuously reduced. From the abovementioned, cinnamic acids naturally exist in various forms and an enzyme reaction system has been established. Therefore, by

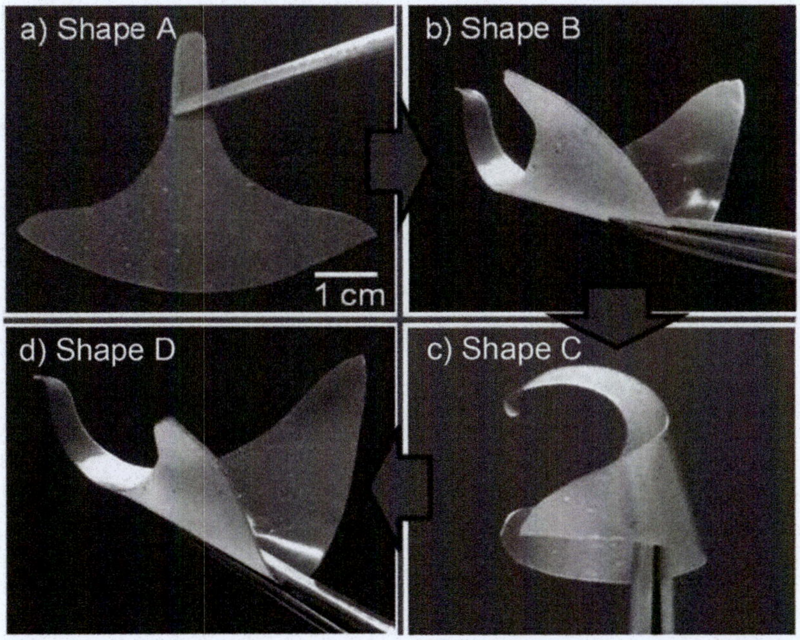

Fig. 13 A typical complex photoinduced shape-memory effect. **a** A film with shape A processed by a hot press. **b** The shape B was processed from shape A by repeating irradiation by UV at 280–450 nm. **c** The shape C was formed at 60 °C and was fixed by cooling to 20 °C while maintaining the mechanical force. **d** The shape D was prepared to put into hot water and then the shape B. Reproduced with permission from *Angew. Chem. Int. Ed.*, **2013**, *52(42)*, 11143–11148. Copyright 2013 Wiley Publishing Group [43]

introducing a gene encoding a required enzyme into a microorganism such as *Escherichia coli* by a gene recombination method. These microorganisms are used as a microreactor to produce the desired cinnamic acid from glucose or the appropriate amino acids. In addition to this enzymatic reaction system, the structure of the antibiotic Pristinamycin I produced by microorganisms (*Streptomyces pristinaespiralis*) contains an aromatic diamine derivative called 4-aminophenylalanine (4APhe) which is obtained by bioproductions [44]. 4APhe was expected to use as raw materials of polyimides because it has aromatics and diamines in its structure. Since 4APhe-derived polyimide has a carboxylic acid in the side chain, it exhibits high solubility in various organic solvents and was dissolved in amide solvents such as DMF and DMSO (Fig. 14) [45].

In addition, depending on the type of counter monomers of tetracarboxylic diacids, it showed the property of being soluble in tetrahydrofuran (THF) and dichloromethane (CH_2Cl_2). In addition, all the polyimides showed extremely high heat resistance, and at a 10% weight loss temperature (T_{d10}), was higher than 300 °C. These results suggested that this is because the benzimide structure in the structure has a strong influence. In addition to the presence of side chain carboxylic acid residues, the high solubility as described above is thought to be due to the irregular structure of 4APhe due to the asymmetrical heterodiamine. As a more characteristic feature, the unique property of the 4APhe-derived polyimide is that it exhibits the behavior of conversion of solubility and improvement of heat resistance by heat decarboxylation (Fig. 15).

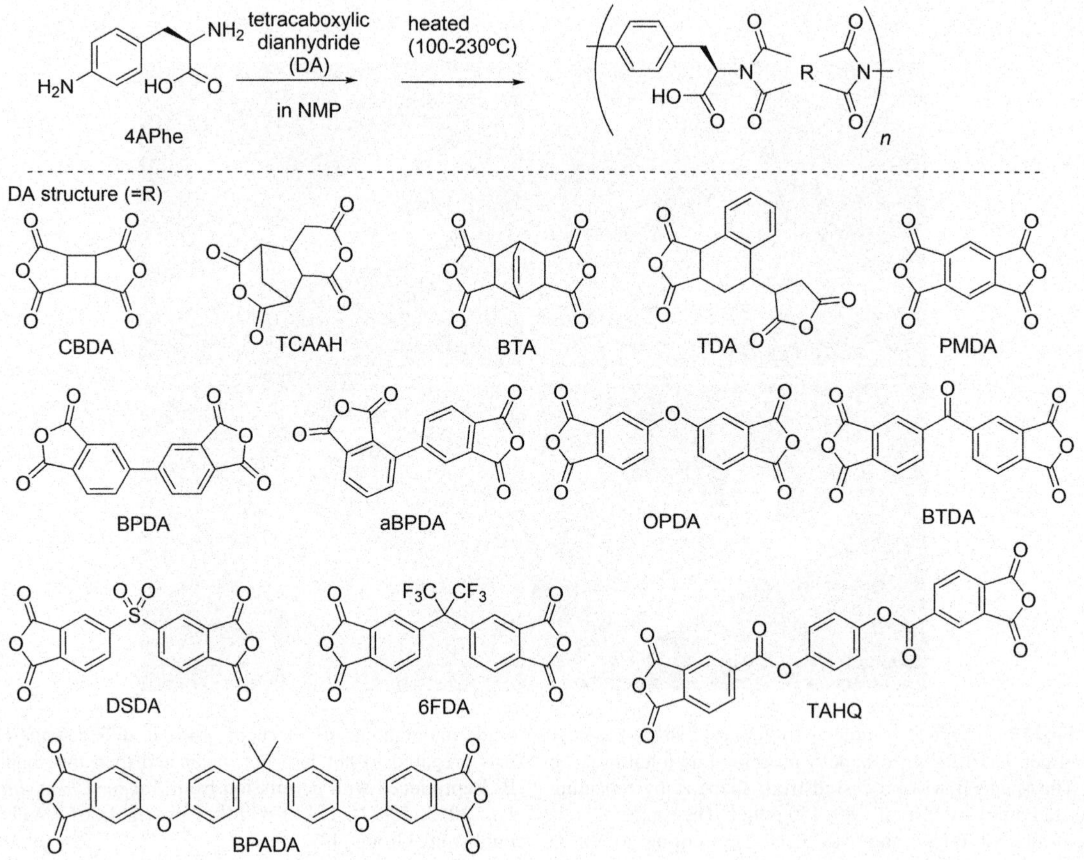

Fig. 14 Syntheses of 4APhe-based polyimides using various tetracarboxylic acid dianhydrides (DAs) [45]

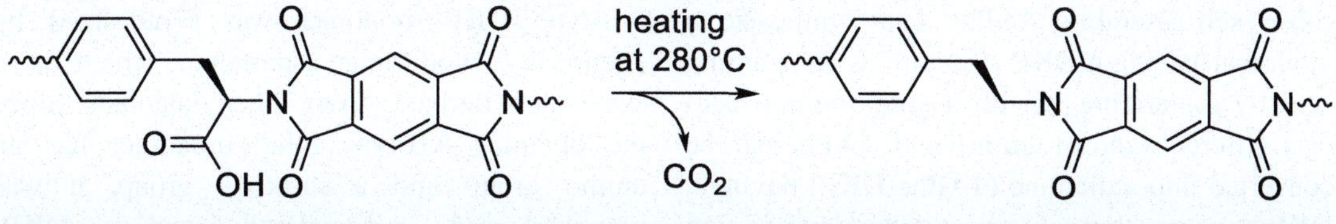

Fig. 15 Mechanism of decarboxylation of 4APhe-based polyimide to improve the heat resistivity and insolubility [45]

Polypeptides form secondary structures through interactions such as hydrogen bonds or π-electron interactions. Ionizable polypeptides such as poly(**L**-lysine) are pH-stimulated responsive, however, there are few studies on cationic aromatic polypeptides. Poly(4-aminophenylalanine) (P4APhe) having a cationic aromatic side chain was synthesized by attempting metal-catalyzed ring-opening polymerization of *N*α-carboxyanhydride using 4APhe as a monomer. The resulting P4APhe had good water solubility in contrast to conventional aromatic poly(amino acids) such as poly(**L**-phenylalanine) or poly(**L**-tyrosine). This P4APhe showed the property of in situ gelation by rehydration in an aqueous solution by forming a crystalline β-sheet domain (Fig. 16). In addition, the cationic copolymer of 4APhe and **L**-lysine exhibited pH response behavior due to changes in the crystalline β-strand, α-helix structure, and random coil balance [46].

3.2 Cyclic Dipeptide-Based Polymers for Self-assembly

Cyclic dipeptides (diketopiperazine, DKP) containing these amino acids have a rigid six-membered ring structure with a *cis*-formed amide bond, and therefore have heat resistance and enzymatic degradation resistance. Also, due to the centrally symmetric cyclic amide bond of DKP, four sites (two N–H and C=O) can form strong intermolecular hydrogen bonds and

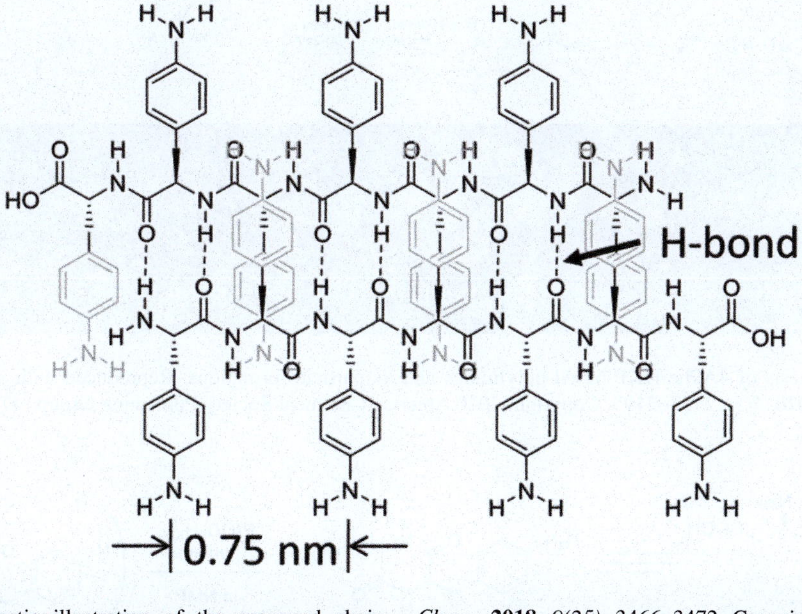

Fig. 16 Schematic illustration of the proposed chain packing of P4APhe based on the formation of β-sheet structures. Reproduced with permission from *Polym. Chem.*, **2018**, *9(25)*, 3466–3472. Copyright 2018 Royal Society of Chemistry Publishing Group [46]

cause self-assembly. 4APhe can synthesize a monomer having a DKP structure as an α-amino acid by dimerizing it with a single amino acid or a different kind of amino acid. 4APhe can be converted into a diamine (4APhe-DKP) having a DKP structure through a protection/deprotection reaction. By reacting this 4APhe-DKP with various tetracarboxylic dianhydrides, DKP-based polyamic acid (DKP-PAA) was quantitatively obtained (Fig. 17) [47]. The obtained DKP-PAA was derived from the DKP unit in the main chain and self-assembled into particles, and its morphology could be controlled by changing the polarity of the solvent in the reprecipitate process. In addition, by carrying out the chemical imidization of this DKP-PAA, the polyimide while maintaining the shape of the particles. Furthermore, all the obtained DKP-based polyimide particles had high heat resistance over 350 °C as 5% weight-loss temperature.

Since the DKP-based polyimide particles derived from 4APhe can change their morphology, the morphology control of the particles can be applied to cosmetics, medical treatment, fillers, and the like. The highly soluble AB-type DKP monomer was synthesized by chemically modifying aspartame, which is a sweetener derived from phenylalanine. Since the obtained AB-type DKP monomer has an amino group and a carboxyl group, it was polycondensed independently, and a DKP-derived polyamide (DKP-PA) was obtained by a conventional polycondensation method of the presence of triphenyl phosphite and pyridine (Fig. 18). For the preparation of the particles, a reprecipitation method with various solvents on the obtained DKP-PA was carried out. As a result, irregular network structures and irregular particles such as elliptical particles, hollow particles, and cubes were observed. By comparing these changes in the structure of the particles with infrared spectroscopic analysis and a chemical calculation method called the Hansen solubility parameter [48–50], the morphology of these particles depends on the strength of hydrogen bonds in each solvent in the precipitation process.

From above, 4APhe was modified into a water-soluble polypeptide and a highly heat-resistant biobased soluble polyimide. Cyclic

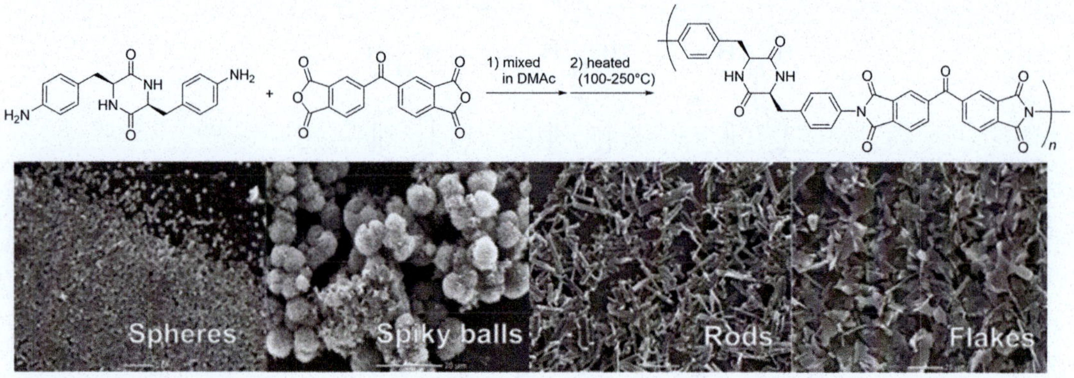

Fig. 17 Synthesis of 4APhe-DKP-based polyimides and its particle formations. Reproduced with permission from *ACS Omega* **2020**, *5(5)*, 2187–2195. Copyright 2020 America Chemical Society Publishing Group [47]

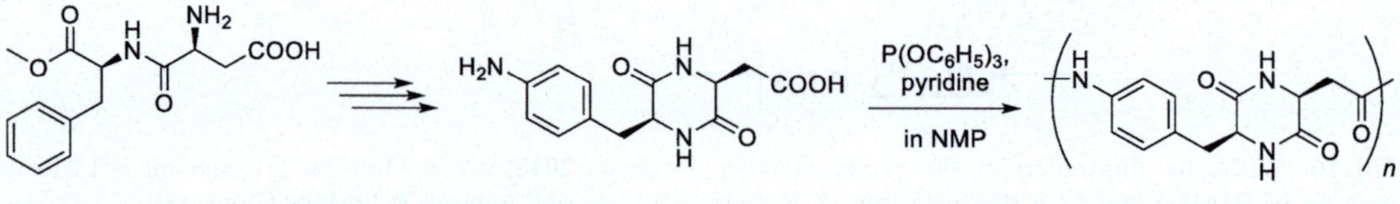

Fig. 18 Synthesis of DKP-PA from aspartame [51]

dipeptide of DKP shows strong intermolecular forces in the polyimide and polyamide, and the polymer forming a unique morphology depends on the solvent due to the main chain DKP unit. Such amino-acid-based functional bioplastics, which able to apply to the development of bioplastics with even higher performance. It is expected that the technology for microbial conversion will continue to develop in the future, and the development of polymers based on amino acids is also expected to show new developments along with this [51].

4 Biobased High-Performance and Functional Polymers Using Cinnamate Dimers

4.1 Biobased Polyimides

Bio-derived 4APhe able to convert to 4-aminocinnamic acid (4ACA) by an enzymatic reaction. It has been studied long decade that cinnamic acid derivatives exhibit *E-Z* (*trans-cis*) isomerization and [2+2] cycloaddition (photodimerization) by UV irradiation [33, 52]. In particular, the photodimerization reaction is a solid-state topochemical reaction from the original crystal state to the product crystal state, and the basic skeleton of the cinnamic acid derivative crystal is retained before and after the reaction. For instance, when the phenylene part is head and the carbonyl part is tail, there are three types of crystal state of cinnamic acid: α type of head–tail (H–T) type packing, β type of head–head (H–H) type packing, and γ type (Fig. 19). The distance between the double bonds is different even in the β type and γ type crystals of the same H–H type packing, and in the γ type crystal, the photodimerization reaction show no active because the distance is too long [53, 54].

By modifying the substituent and crystal state of the cinnamic acid derivatives, it is theoretically possible to obtain a cinnamate dimer having an objective structure. In case of 4ACA, by chemically modifying and UV irradiating to obtain a 4ACA dimer as a diamine, which is a raw material for polyimide. The 4ACA-dimer was reacted with various tetracarboxylic dianhydrides to obtain the poly(amic acid)s and then polyimide through a thermal imidization process (Fig. 20). Notable physical properties of the obtained biobased polyimides are that T_{d10} showed a maximum of 425 °C and a minimum of 390 °C. In addition,

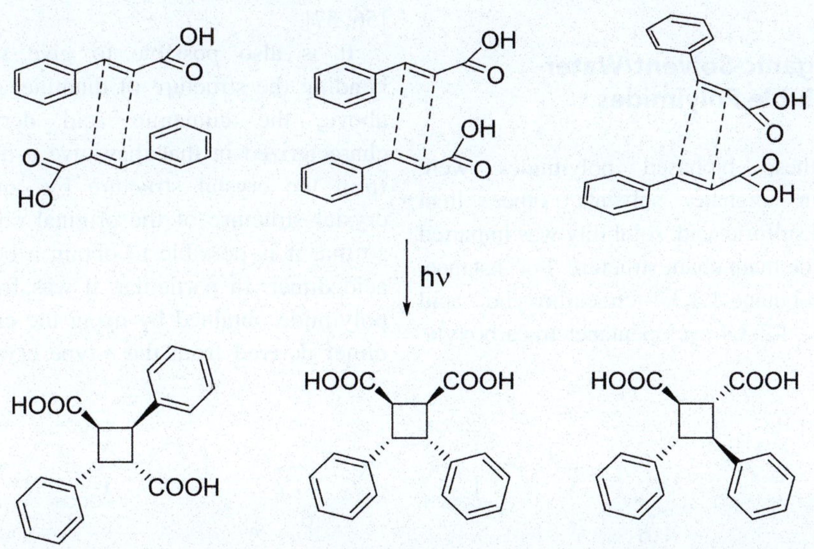

Fig. 19 Formation of the crystalline structure of cinnamic acids and stereostructure of dimers

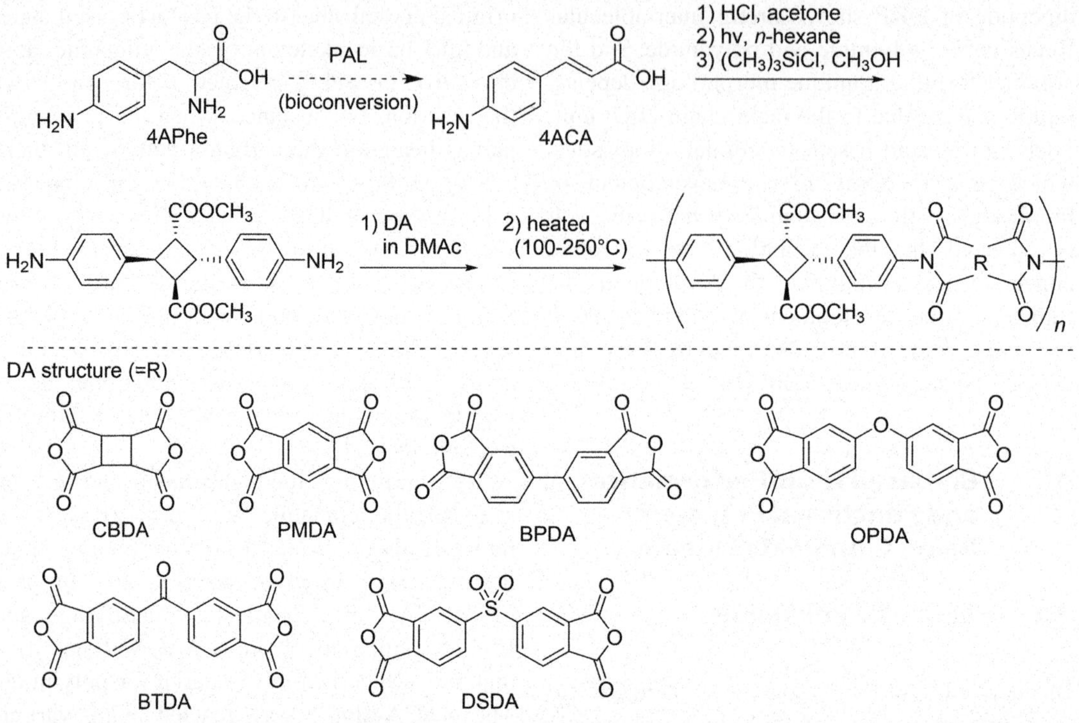

Fig. 20 Synthesis of biobased polyimides using 4ACA-dimer derived from 4ACA [55]

the polyimide obtained by using 1,2,3,4-cyclobutanetetracarboxylic dianhydride, which is an aliphatic tetracarboxylic dianhydride, has a lower yellowness and higher transparency than Kapton™. It showed a transmittance of about 90% at a wavelength of 450 nm [55].

4.2 Organic-Solvent/Water-Soluble Polyimides

Although these biobased polyimides were insoluble in organic solvents other than concentrated sulfuric acid, solubility was imparted by bending the main chain structure. For instance, using meso-butane-1,2,3,4-tetracarboxylic acid dianhydride, 1,2,3,4-cyclopentanetetracarboxylic dianhydride, and 1,2,4,5-cyclohexanetetracarboxylic acid dianhydride, which are alicyclic bending monomers, the soluble polyimides were obtained (Fig. 21). These showed the solubility in organic solvents (NMP, DMSO, DMF, DMAc) while maintaining the high heat resistance of polyimide [56, 57].

It is also possible to give solubility by bending the structure of diamine. As described above, the cinnamic acid derivatives are characterized in that they give a dimer derived from the crystal structure by controlling the crystal structure of the original cinnamic acid, so that it is possible to obtain a bent cinnamic acid dimer. In particular, it was found that the polyimide obtained by using the cinnamic acid dimer derived from the δ-type crystal structure

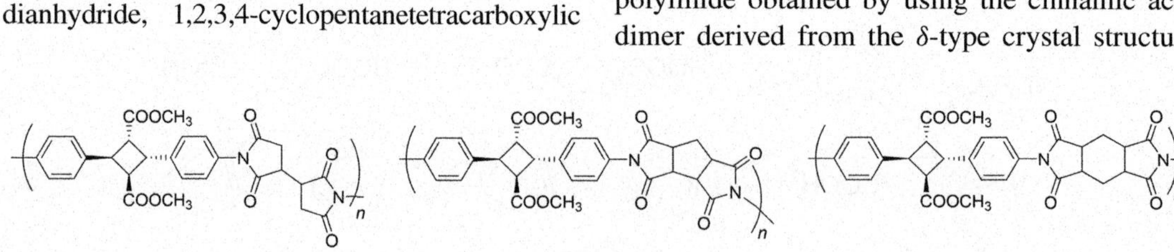

Fig. 21 Structure of soluble biobased polyimides using bending tetracarboxylic acid dianhydrides [56, 57]

as a diamine shows good solubility in an organic solvent [58]. With such a bendable polyimide, it is possible to recycle the molded polyimide film by dissolving it with a solvent and re-forming it into a film (Fig. 22).

4ACA-derived polyimide has a reactive carboxylic acid in the side chain. By treating the side chain with an aqueous alkali metal solution or the like by utilizing this property, it is also possible to ionize the side chain carboxylic acid and make it water-soluble. Since water-soluble polymers dissolve in water, which is the safest and relatively easy-to-remove solvent, they can be handled in a low-environmental manner and film can be prepared in an environment without intake drafts. Furthermore, it is also possible to composite with other water-soluble materials (cellulose nanofibers [59–62] and polyacrylates [63–66], etc. [67]). The 4ACA-based polyimide having a carboxylic acid in the side chain of polyimide that was insoluble in organic solvents and water. This is because the regular arrangement of carboxyl groups enhances the interaction between each polyimide molecule. However, by immersing this polyimide film in an aqueous alkali metal solution such as sodium hydroxide or potassium hydroxide, the film is completely dissolved in water (Fig. 23) [68].

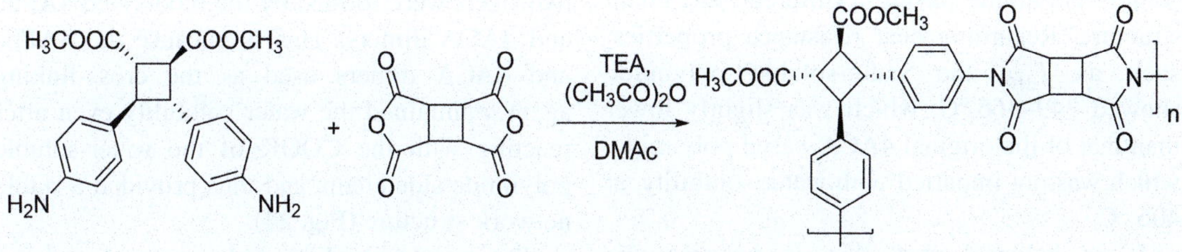

Fig. 22 Synthesis of soluble biobased polyimides using bending 4ACA-dimer [58]

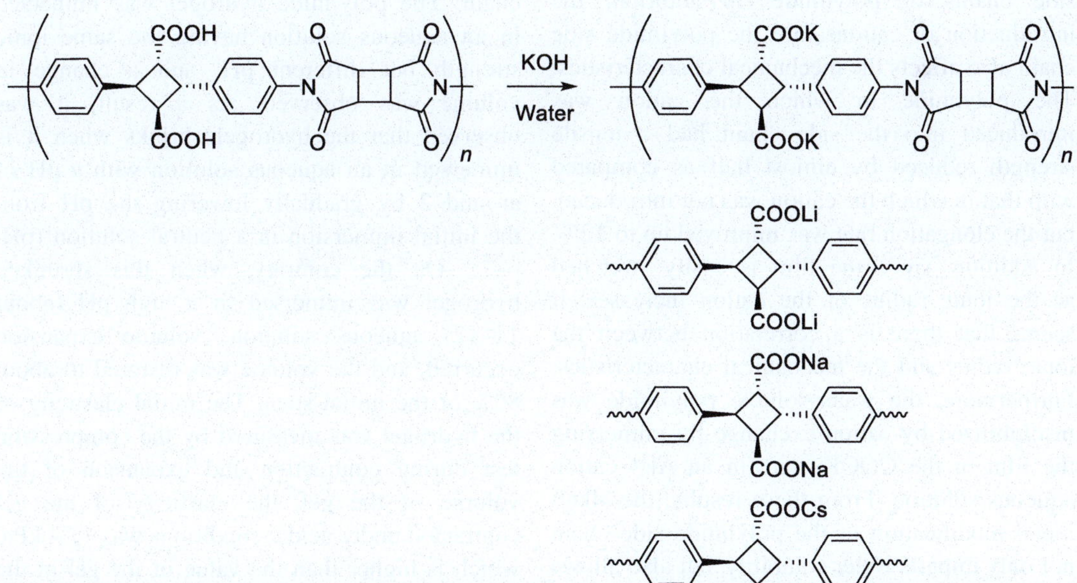

Fig. 23 Preparation of water-soluble biobased polyimidebiobased with Li, Na, K, and Cs salt in the side chain [68]

When this aqueous solution was added dropwise to ethanol, the powder derived from the polymer was reprecipitated, and when the precipitate was measured by ^{1}H NMR, it was found to have a 4ACA-based polyimide structure. In addition, the obtained water-soluble polyimide maintained a high molecular weight at a level that could be molded as a film. From the results of IR measurement, it was confirmed that the polyimide was not decomposed by the treatment with the alkali metal aqueous solution because the absorption of the amide bond derived from the precursor poly(amic acid) was not observed. By using LiOH, NaOH, and CsOH by such a method, it was possible to obtain the water-soluble polyimides in the same manner, and it was possible to obtain a water-soluble polyimide having a different side chain structure. Regarding heat resistance properties, such as T_{d10}, the water-soluble polyimide showed 349–366 °C, which was slightly lower than that of the original 4ACA-based polyimide, which was not imparted with water solubility, at 406 °C.

It is believed that this is because the introduction of alkali metal-derived cations weakened the interaction between the side chains of polyimide. In addition, the introduction of cations into the polyimide side chain also affects the mechanical characteristics. The polyimide in which the cation was introduced into the side chain had a tensile strength reduced by almost half as compared with that in which the cation was not introduced, but the elongation rate was improved up to 12%. In addition, since the film is easily stretched as the ionic radius of the cation increases, it seems that there is a correlation between the ionic radius and the mechanical characteristics. Furthermore, the water-soluble polyimide was insolubilized by cation exchange by immersing the film in the COOK state in an Al^{3+} cation aqueous solution. From these results, the alkali metal modification of the polyimide side chain not only imparts water solubility, but also allows the physical properties to be freely controlled to some extent by the cation species.

4.3 Polyimide Hydrogels for PH-Responsive Gel

Based on the result that the side chain can be substituted with cations, it is possible to synthesize a polyimide hydrogel by applying the reaction in an aqueous solution [69]. 4-(4,6-Dimethoxy-1,3,5-triazine-2-yl)-4-methylmorpholinium chloride (DMT-MM) is a condensing agent capable of carrying out a condensation reaction between a carboxylic acid and an amino group in water, and a hydrogel is formed by reacting a water-soluble polyimide with various diamines in the presence of DMT-MM. The diamines were used alkyl chain and the aromatic amino acids 4APhe and 4ACA dimers, and it happened that the hydrogels were formed by the bioderived 4APhe and 4ACA dimers. This is because the 4APhe and 4ACA dimers used as the cross-linking agent maintained the water solubility even after reacting with the COOK of the water-soluble polyimide side chain, and thus provided a stable network structure (Fig. 24).

The polyimide hydrogel causes a volume change due to a change in intramolecular interaction due to cation exchange in the side chain. The polyimide hydrogel was immersed in an aqueous solution having the same ionic strength but different pH, and a change in volume was observed. As a result, it was observed that the hydrogel shrinks when it is immersed in an aqueous solution with a pH of around 2 by gradually lowering the pH from the initial immersion in a neutral solution (pH, 6–7). On the contrary, when this shrunken hydrogel was immersed in a high pH (about 11–12) aqueous solution, volume expansion occurred, and the volume was restored to about 80% of the initial state. The initial elasticity of the hydrogel was measured by the compression test during contraction and expansion of the volume of the gel, the elasticity of the gel contracted under acidic conditions was 17.4 kPa, which is higher than the value of the gel in the initial state (6.7 kPa). The swollen gels under alkaline conditions showed almost the same

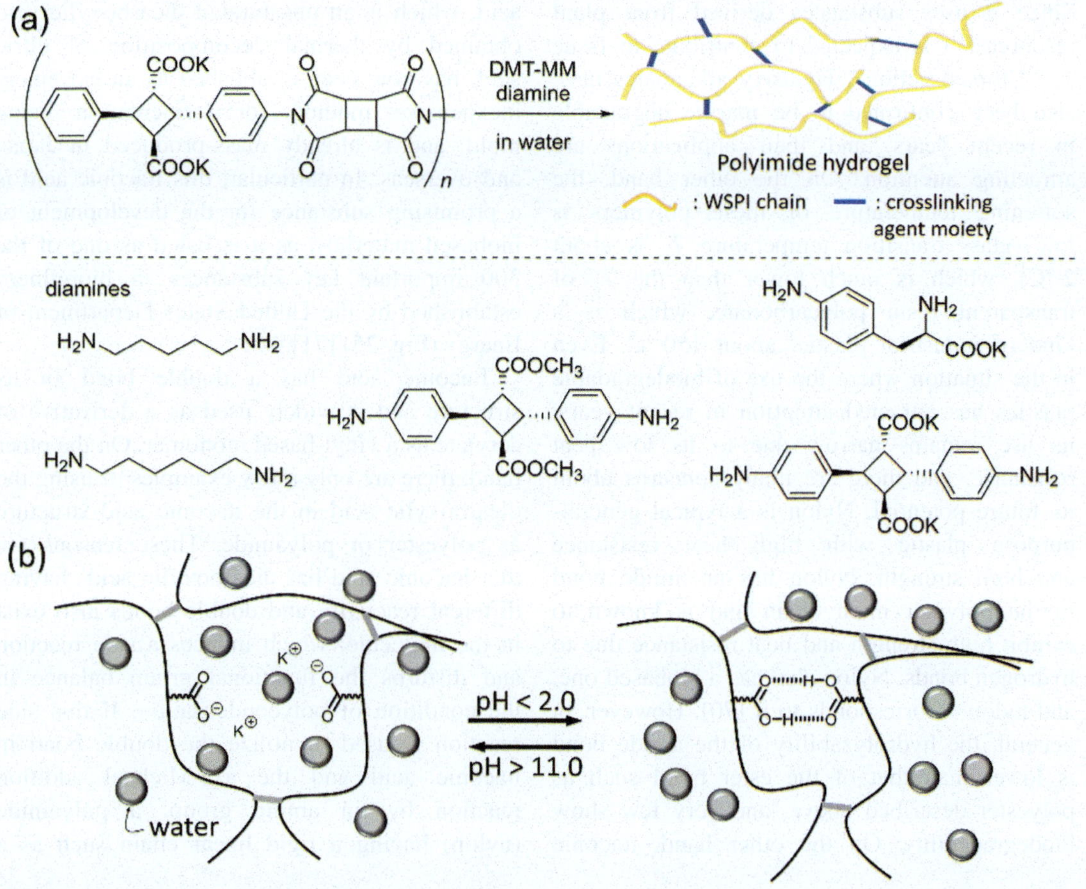

Fig. 24 **a** Preparation of biobased polyimide hydrogel by crosslinking reaction using various diamines and DMT-MM as crosslinking agents and condensation agents. **b** Contract and swelling mechanism of polyimide hydrogel [69]

elasticity (7.8 kPa) as the initial gels before pH adjustment, and the deformation of these gels changed in density.

In other words, it was found that the volume changed due to the ingress and egress of water molecules. Similar to the poly(acrylic acid)-based pH-responsive hydrogel, this phenomenon of volume change is said to affect the volume change of the hydrogel by changing the intermolecular interaction due to the exchange of protons accompanying the pH change. In particular, the polyimide hydrogel is completely bio-derived, it was expected to be used in a drug delivery system in the body, which is one of the uses of pH-responsive gels. In addition, since these 4ACA-based polyimides and polyamides have a cyclobutane ring in the main chain, they cause decomposition of the main chain by irradiating them with UV of a short wavelength, which is possible to recycle a raw material of 4ACA.

5 Itaconic Acid-Based Polyamides as a Degradable Plastics

Plastics and the environment are closely related fields as abovementioned, and in the Sustainable Development Goals (SDGs), not only Issue 12 "Responsible consumption, production" but also Issue 9 "Industry, innovation, infrastructure".

Since it uses substances derived from plant resources, it is expected to contribute to Issue 13 "Climate action". Hydroxy alkanoates have also been confirmed to be marine degradable in recent years, and their applications are attracting attention. On the other hand, the softening temperature of these polymers is low (glass transition temperature T_g is about 2 °C), which is much lower than the T_g of transparent resin polycarbonate, which is a kind of industrial plastic, about 150 °C. Even in the situation where the use of biodegradable plastics has attracted attention in recent years, its use remains narrow due to its low heat resistance, and there are many concerns about its future potential. Nylon is a typical general-purpose plastic with high heat resistance and high strength. Nylon has an amide bond in the polymer main chain and is known to exhibit high strength and heat resistance due to hydrogen bonds. Nylon also has a biobased one, and nylon 4 corresponds to it [70]. However, in general, the hydrolyzability of the amide bond is lower than that of the ester bond such as polyester described above, and very few show biodegradability. On the other hand, itaconic acid, which is an unsaturated dicarboxylic acid obtained by thermal decomposition of citric acid, has now been established as an extremely inexpensive method for fermentation using mold, and is already mass-produced in Japan and overseas. In particular, this itaconic acid is a promising substance for the development of biobased materials, as it is listed as one of the 300 important key substances in biorefinery established by the United States Department of Energy (Fig. 25) [71].

Itaconic acid has a double bond in its structure and is widely used as a derivative of acrylate as a vinyl-based monomer. On the other hand, there are only a few examples of using the dicarboxylic acid in the itaconic acid structure as polyester or polyamide. These reasons are that itaconic acid has dicarboxylic acids having different reactivity and double bonds also exist in the molecule, which induces a side reaction and disturbs the functional group balance in the condition of polycondensation. If this side reaction is used to utilize the double bond in itaconic acid and the aza-Michael addition reaction by an amino group, a polyamide (nylon) having a rigid linear chain such as a

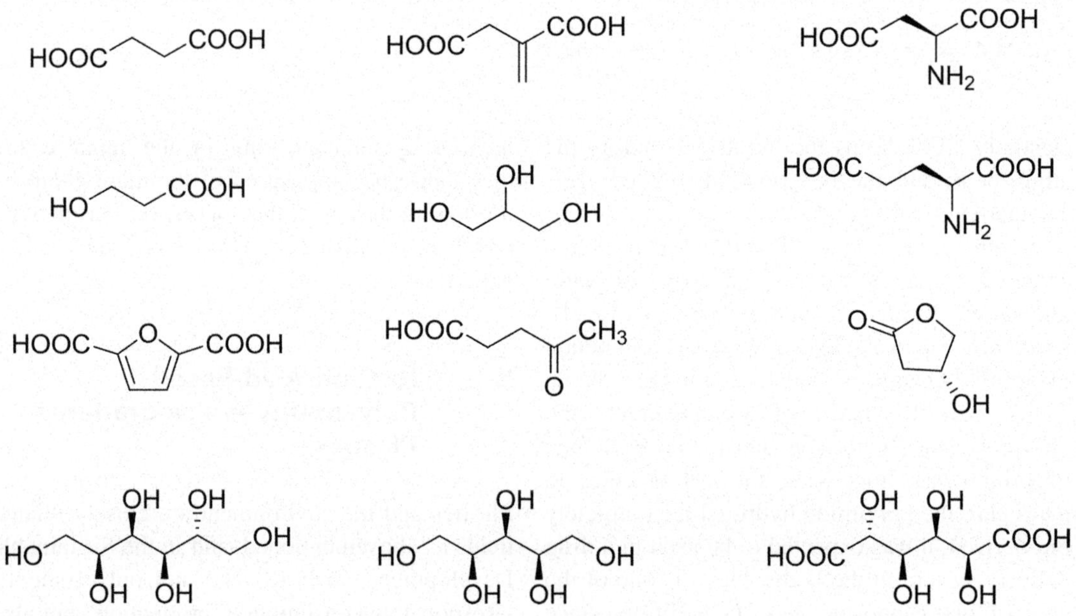

Fig. 25 Major biomolecules proposed by the US Department of Energy [71]

pyrrolidone ring in the main chain due to the subsequent dehydration reaction [72, 73]. In order to achieve such a molecular design, the aza-Michael addition reaction must first occur, and then the carboxylic acid of itaconic acid and the amine must be dehydrated and condensed to form a five-membered ring. However, if the reaction between the carboxylic acid and the amine occurs first, a double bond remains, and another amine and aza-Michael addition reaction occurs there, resulting in the loss of the 1:1 balance of functional groups (Fig. 26).

Therefore, it causes the difficulty to obtain a high molecular weight polyamide. In order to solve this, it is effective to make a nylon salt from itaconic acid and diamine and heat bulk polymerization using it as a salt monomer. As a result, the functional group balance between the carboxylic acid and the diamine at the reaction site is restricted to 1:1 and the amine becomes a salt state, so that the reactivity is suppressed. Such a complicated reaction system is analyzed by thermogravimetric analysis, and the curve obtained by differential thermal analysis (DTA) draws a curve in which two endotherms and one heat generation are combined.

This means that the amino group causes an aza-Michael addition reaction to the double bond with the carbonyl group of itaconic acid, gives a secondary amine, and immediately undergoes a dehydration condensation reaction with the carboxylic acid at high temperature. Since winding a 5-membered ring produces a stable structure rather than winding a 4-membered ring, a pyrrolidone ring is formed by condensing the remaining carboxylic acid with an amine.

The T_g of nylon using an aliphatic diamine was 80–97 °C, and the T_{d10} was 300–410 °C. It was found that T_g and T_{d10} of nylon using aromatic diamine were 156–242 °C and 370–400 °C, respectively, and had sufficient heat resistance as an engineering plastic. T_g is higher than the conventional nylon 6,6 and nylon 6 at 57 and 53 °C, and was used under higher temperature conditions. On the other hand, T_{d10} was almost equivalent to these conventional nylons (390 and 400 °C). It is considered that the above high heat resistance is caused by the fact that one amide bond retains a hydrogen bond, while the other amide forms a pyrrolidone ring and imparts rigidity to the polymer main chain. The mechanical properties of these aliphatic nylons were a breaking strength of 65–90 MPa and a Young's modulus of 430–2800 MPa. These are also values that exceed the mechanical properties of nylon 6,6 (breaking strength: 65–68 MPa, Young's modulus: 590–1700 MPa) and nylon 6 (mechanical strength: 41–60 MPa, Young's modulus: 440–1400 MPa) [74–77]. Aromatic nylon is derived from the pyrrolidone ring and aromatic, and the main chain structure has high rigidity, which causes brittle properties as a material. Therefore, polyamides using itaconic acid and aliphatic dicarboxylic acids are promising to be used as a nylon fiber.

Further, since the bio-based nylons have a pyrrolidone ring in the main chain, it is possible

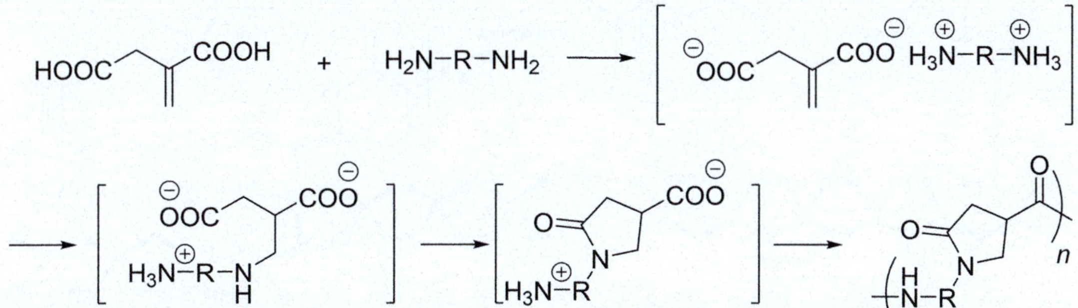

Fig. 26 Schematic illustration of the formation of pyrrolidone ring and synthesis of polyamide by aza-Michael addition reaction via salt monomer [72, 73]

to induce these ring-opening reactions by some kind of stimulation. Cyclic amides such as the pyrrolidone ring show a ring-opening reaction without inducing cleavage of the main chain by hydrolysis, and thus can cause degradability or softening of the material. Nylon derived from itaconic acid showed the property of dissolving to water under alkaline conditions. This property is also exhibited in soil showing weak alkalinity (pH 7.5–7.9), and when nylon was buried to a depth of 2–4 cm and its decomposition test was carried out. 1,3-Diaminopropane, 1,4-diaminobutane, 1,5-diaminopentane, 1,6-diaminohexane-based nylon synthesized with aliphatic diamines were disappeared in the soil. It is considered that the above results disappeared in the soil due to the decomposition under alkaline conditions and the biological action caused by the long-term exposure to the environment, and the improvement of the water solubility of nylon. Furthermore, by applying such a photo-ring-opening reaction of the pyrrolidone ring, it is possible to decompose the nylon resin in water under UV irradiation. The fact that it can be decomposed in water by UV suggests that it can also be decomposed by sunlight at sea, and it will contribute to the reduction of the recent marine plastic waste problem. It is also possible to make nylon that decomposes with pepsin, which is a digestive enzyme, by using amino acids for diamine, which is a counter monomer of itaconic acid [78]. The fact that the resin decomposes and disappears in the UV and the digestive tract is an important performance in achieving the SDGs target.

It has been known that nylon can greatly improve its mechanical strength by compositing with a clay mineral called montmorillonite (MMT) [79–83]. Bionylon obtained using itaconic acid also be composited with MMT. Since bionylon melts by heating, it is possible to modify the mechanical properties of bionylon by mixing the melted bionylon and MMT and subjecting them to ultra-low temperature treatment with liquid nitrogen [84, 85]. From this result, a strong and ductile bionylon/MMT composite was obtained [86]. The tensile strength and Young's modulus of the prepared composite were 3.2–5.5 GPa and 65–241 MPa, respectively, which were equal to or higher than those of the conventional glass fiber reinforced nylon 6. The bionylon/NaMMT hybrid with a clay content of 7 wt% showed the highest elongation at break and its strain energy density reached 21 MJ m^{-3}. The reason for this improvement in elongation is that the NaMMT catalyzes the ring-opening of the pyrrolidone ring in the bionylon backbone to form alkoxylic acid, which gives flexibility to the polymer structure (Fig. 27). It is considered that this

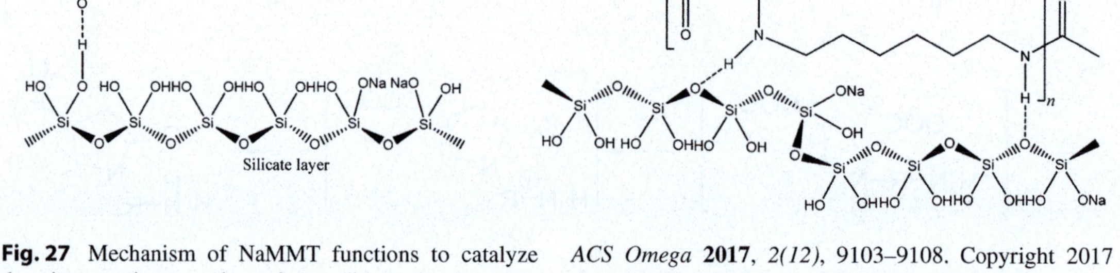

Fig. 27 Mechanism of NaMMT functions to catalyze the ring-opening reaction of pyrrolidone in the main chain of bionylon. Reproduced with permission from *ACS Omega* **2017**, *2(12)*, 9103–9108. Copyright 2017 America Chemical Society Publishing Group [86]

is because the strong interaction between the silicate layer and the polymer chain was induced.

As described above, the development of a heterocyclic bionylon was achieved by using itaconic acid produced by microorganisms. Pyrrolidone rings are formed in the nylon chain by performing melt polycondensation via salt monomers with itaconic acid and various diamines, and the pyrrolidone rings exhibit unique properties, which are alkaline soil and UV-responsive degradability was imparted to nylon. It is considered that these properties were adjusted by combining the composition of nylon salt, and it is expected to be applied to degradable switching in the environment. As an example, if bionylon is used for fishing line, even if it is dumped in the ocean during fishing, it will gradually deteriorate (lower molecular weight) from the UV rays contained in sunlight and become water-soluble, thereby promoting biodegradation. It is expected to be used as an ideal degradable plastic. In addition, the bionylon structure was modified by amino acid to produce resin collapsed in digestive enzymes such as pepsin [78]. The amino acid used for this was d- or **L**-leucine. Chiral dicarboxylic acid monomers derived from renewable itaconic acid and leucines were synthesized with high purity. Chiral interactive polyamides based on **D**- or **L**-leucine were prepared via melt polycondensation of hexamethylenediamine. The chiral-interacting polymer composites showed T_g of approximately 117 °C and T_m, of approximately 213 °C, which are higher than those of conventional bioderived polyamides such as polyamide 11 (T_g of approximately 57 °C). In addition, the bionylon resin collapsed in the presence of a digestive enzyme, pepsin, which suggests that the resin should not be stacked in the stomach and around the digestive tract even if marine animals mistake swallowing. In this way, the bionylons will be important as a topic that provides clues for solving the plastic waste problem in recent years.

6 Summary

The various polymers can be obtained by using amino acids produced by microorganisms using glucose or amino acids that are metabolically designed by gene recombination. In addition, these polymers have unique properties not found in conventional petroleum-based polymers and are unique because they are biobased unique structures. Particularly in recent years, there has been a demand for biobased materials and the development of degradable plastics that have not remained in the environment for a long time as garbage. In order to tackle these problems, various biobased polymers and degradable plastics are still being developed, and researches on the degradability of plastics based on microbiology is still being conducted. Expectations are rising for the development of various materials.

Acknowledgements Theses research are based on the results of the New Energy and Industrial Technology Development Organization (NEDO)—MOONSHOT Research and Development Program—"JPNP18016" and the Japan Society for the Promotion of Science of Grant-in-Aid for Early-Career Scientists (21K14681). In addition, the development of the pathway for microbial production of 4ACA is supported by our research collaborator Professor Naoki Takaya (University of Tsukuba) and the staff of his laboratory.

References

1. E.S. Beach, Z. Cui, P.T. Anastas, Green chemistry: a design framework for sustainability. Energy Environ. Sci. **2**(10) (2009). https://doi.org/10.1039/b904997p
2. S.A. Miller, Sustainable polymers: opportunities for the next decade. ACS Macro Lett. **2**(6), 550–554 (2013). https://doi.org/10.1021/mz400207g
3. P.A. Wilbon, F. Chu, C. Tang, Progress in renewable polymers from natural terpenes, terpenoids, and rosin. Macromol. Rapid Commun. **34**(1), 8–37 (2013). https://doi.org/10.1002/marc.201200513
4. V. Froidevaux, C. Negrell, S. Caillol, J.P. Pascault, B. Boutevin, Biobased amines: from synthesis to polymers; present and future. Chem. Rev. **116**(22), 14181–14224 (2016). https://doi.org/10.1021/acs.chemrev.6b00486

5. C.J. Clarke, W.C. Tu, O. Levers, A. Brohl, J.P. Hallett, Green and sustainable solvents in chemical processes. Chem. Rev. **118**(2), 747–800 (2018). https://doi.org/10.1021/acs.chemrev.7b00571
6. J. Rosler, F. Krekel, N. Amrhein, J. Schmid, Maize phenylalanine ammonia-lyase has tyrosine ammonia-lyase activity. Plant Physiol. **113**(1), 175–179 (1997). https://doi.org/10.1104/pp.113.1.175. Accessed 6 Jan 2022
7. A.M. Patten, M. Jourdes, E.E. Brown, M.-P. Laborie, L.B. Davin, N.G. Lewis, Reaction tissue formation and stem tensile modulus properties in wild-type and p-coumarate-3-hydroxylase downregulated lines of alfalfa, Medicago sativa (Fabaceae). Am. J. Bot. **94**(6), 912–925 (2007). https://doi.org/10.3732/ajb.94.6.912
8. D. Schomburg, I. Schomburg, A. Chang (eds.), 2-Hydroxypropyl-CoM lyase, in *Class 4–6 Lyases, Isomerases, Ligases: EC 4–6* (Springer Berlin Heidelberg, 2009), pp. 407–410
9. J.J. Stewart, T. Akiyama, C. Chapple, J. Ralph, S.D. Mansfield, The effects on lignin structure of overexpression of ferulate 5-hydroxylase in hybrid poplar. Plant Physiol. **150**(2), 621–635 (2009). https://doi.org/10.1104/pp.109.137059. Accessed 6 Jan 2022
10. K.J. Hellingwerf, J. Hendriks, T. Gensch, Photoactive yellow protein, a new type of photoreceptor protein: will this "yellow lab" bring us where we want to go? J. Phys. Chem. A **107**(8), 1082–1094 (2003). https://doi.org/10.1021/jp027005y
11. J. Tenboer, S. Basu, N. Zatsepin, K. Pande, D. Milathianaki, M. Frank, M. Hunter, S. Boutet, G.J. Williams, J.E. Koglin et al., Time-resolved serial crystallography captures high-resolution intermediates of photoactive yellow protein. Science **346**(6214), 1242–1246 (2014). https://doi.org/10.1126/science.1259357
12. X. Jin, C. Carfagna, L. Nicolais, R. Lanzetta, Synthesis, characterization, and in vitro degradation of a novel thermotropic ternary copolyester based on p-hydroxybenzoic acid, glycolic acid, and p-hydroxycinnamic acid. Macromolecules **28**(14), 4785–4794 (1995). https://doi.org/10.1021/ma00118a001
13. J. Stumpe, A. Ziegler, M. Berghahn, H.R. Kricheldorf, Photoreactive cholesteric polyesters derived from 4-carboxycinnamic acid and novel chiral spacers. Macromolecules **28**(15), 5306–5311 (1995). https://doi.org/10.1021/ma00119a021
14. B. Sapich, J. Stumpe, T. Krawinkel, H.R. Kricheldorf, New polymer syntheses. 95. Photosetting cholesteric polyesters derived from 4-hydroxycinnamic acid and isosorbide. Macromolecules **31**(4), 1016–1023 (1998). https://doi.org/10.1021/ma971027e
15. T. Kaneko, M. Matsusaki, T.T. Hang, M. Akashi, Thermotropic liquid-crystalline polymer derived from natural cinnamoyl biomonomers. Macromol. Rapid Commun. **25**(5), 673–677 (2004). https://doi.org/10.1002/marc.200300143
16. T. Kaneko, D. Kaneko, S. Wang, High-performance lignin-mimetic polyesters. Plant Biotechnol. **27**(3), 243–250 (2010). https://doi.org/10.5511/plantbiotechnology.27.243
17. T. Kaneko, T.H. Thi, D.J. Shi, M. Akashi, Environmentally degradable, high-performance thermoplastics from phenolic phytomonomers. Nat. Mater. **5**(12), 966–970 (2006). https://doi.org/10.1038/nmat1778
18. M. Chauzar, S. Tateyama, T. Ishikura, K. Matsumoto, D. Kaneko, K. Ebitani, T. Kaneko, Hydrotalcites catalyze the acidolysis polymerization of phenolic acid to create highly heat-resistant bioplastics. Adv. Func. Mater. **22**(16), 3438–3444 (2012). https://doi.org/10.1002/adfm.201200427
19. D. Kaneko, K. Matsumoto, S. Kinugawa, S. Tateyama, T. Kaneko, Effects of adhesive characteristics of the catechol group on fiber-reinforced plastics. Polym. J. **43**(11), 944–947 (2011). https://doi.org/10.1038/pj.2011.91
20. D. Kaneko, S. Wang, K. Matsumoto, S. Kinugawa, K. Yasaki, D.H. Chi, T. Kaneko, Mussel-mimetic strong adhesive resin from bio-base polycoumarates. Polym. J. **43**(10), 855–858 (2011). https://doi.org/10.1038/pj.2011.77
21. S. Wang, D. Kaneko, K. Kan, X. Jin, T. Kaneko, Syntheses of hyperbranched liquid-crystalline biopolymers with strong adhesion from phenolic phytomonomers. Pure Appl. Chem. **84**(12), 2559–2568 (2012). https://doi.org/10.1351/PAC-CON-12-05-12
22. S. Wang, Y. Kitamura, N. Hiraishi, S. Taira, A. Tsuge, T. Kaneko, D. Kaneko, Preparation of mussel-inspired biopolyester adhesive and comparative study of effects of meta- or para-hydroxyphenylpropionic acid segments on their properties. Polymer **165**, 152–162 (2019). https://doi.org/10.1016/j.polymer.2019.01.012
23. J. Kim, J.W. Kim, H.C. Kim, L. Zhai, H.-U. Ko, R.M. Muthoka, Review of soft actuator materials. Int. J. Precis. Eng. Manuf. **20**(12), 2221–2241 (2019). https://doi.org/10.1007/s12541-019-00255-1
24. A. Ryabchun, A. Bobrovsky, J. Stumpe, V. Shibaev, Novel generation of liquid crystalline photo-actuators based on stretched porous polyethylene films. Macromol. Rapid Commun. **33**(11), 991–997 (2012). https://doi.org/10.1002/marc.201100837
25. Z. Wen, K. Yang, J.-M. Raquez, A review on liquid crystal polymers in free-standing reversible shape memory materials. Molecules **25**(5), 1241 (2020)
26. Y. Osada, J.-P. Gong, Soft and wet materials: polymer gels. Adv. Mater. **10**(11), 827–837 (1998). https://doi.org/10.1002/(SICI)1521-4095(199808)10:11<827::AID-ADMA827>3.0.CO;2-L
27. A. Lendlein, H. Jiang, O. Jünger, R. Langer, Light-induced shape-memory polymers. Nature

434(7035), 879–882 (2005). https://doi.org/10.1038/nature03496
28. H. Yu, T. Ikeda, Photocontrollable liquid-crystalline actuators. Adv. Mater. **23**(19), 2149–2180 (2011). https://doi.org/10.1002/adma.201100131
29. M. Yamada, M. Kondo, J.-I. Mamiya, Y. Yu, M. Kinoshita, C.J. Barrett, T. Ikeda, Photomobile polymer materials: towards light-driven plastic motors. Angew. Chem. Int. Ed. **47**(27), 4986–4988 (2008). https://doi.org/10.1002/anie.200800760
30. Y. Yu, M. Nakano, T. Ikeda, Directed bending of a polymer film by light. Nature **425**(6954), 145–145 (2003). https://doi.org/10.1038/425145a
31. H.K. Bisoyi, Q. Li, Light-driven liquid crystalline materials: from photo-induced phase transitions and property modulations to applications. Chem. Rev. **116**(24), 15089–15166 (2016). https://doi.org/10.1021/acs.chemrev.6b00415
32. M. Nagarathinam, A.M. Peedikakkal, J.J. Vittal, Stacking of double bonds for photochemical [2+2] cycloaddition reactions in the solid state. Chem. Commun. (Camb.) **42**, 5277–5288 (2008). https://doi.org/10.1039/b809136f
33. J.R. Mallette, J.F. Casale, Rapid determination of the isomeric truxillines in illicit cocaine via capillary gas chromatography/flame ionization detection and their use and implication in the determination of cocaine origin and trafficking routes. J. Chromatogr. A **1364**, 234–240 (2014). https://doi.org/10.1016/j.chroma.2014.08.072
34. J.B. Metternich, R. Gilmour, A bio-inspired, catalytic E –> Z isomerization of activated olefins. J. Am. Chem. Soc. **137**(35), 11254–11257 (2015). https://doi.org/10.1021/jacs.5b07136
35. S. Wang, S. Tateyama, D. Kaneko, S.-Y. Ohki, T. Kaneko, Synthesis of well-defined hyperbranched polymers bio-based on multifunctional phenolic acids and their structure–thermal property relationships. Polym. Degrad. Stab. **96**(12), 2048–2054 (2011). https://doi.org/10.1016/j.polymdegradstab.2011.10.003
36. T.H. Thi, M. Matsusaki, D. Shi, T. Kaneko, M. Akashi, Synthesis and properties of coumaric acid derivative homo-polymers. J. Biomater. Sci. Polym. Ed. **19**(1), 75–85 (2008). https://doi.org/10.1163/156856208783227668
37. K. Takada, K. Yasaki, S. Rawat, K. Okeyoshi, A. Kumar, H. Murata, T. Kaneko, Photoexpansion of biobased polyesters: mechanism analysis by time-resolved measurements of an amorphous polycinnamate hard film. ACS Appl. Mater. Interfaces **13**(12), 14569–14576 (2021). https://doi.org/10.1021/acsami.0c22922
38. K. Yasaki, T. Suzuki, K. Yazawa, D. Kaneko, T. Kaneko, Effects of double photoreactions on polycoumarate photomechanics. J. Polym. Sci., Part A: Polym. Chem. **49**(5), 1112–1118 (2011). https://doi.org/10.1002/pola.24525
39. T. Hugel, N.B. Holland, A. Cattani, L. Moroder, M. Seitz, H.E. Gaub, Single-molecule optomechanical cycle. Science **296**(5570), 1103–1106 (2002). https://doi.org/10.1126/science.1069856
40. N.B. Holland, T. Hugel, G. Neuert, A. Cattani-Scholz, C. Renner, D. Oesterhelt, L. Moroder, M. Seitz, H.E. Gaub, Single molecule force spectroscopy of azobenzene polymers: switching elasticity of single photochromic macromolecules. Macromolecules **36**(6), 2015–2023 (2003). https://doi.org/10.1021/ma021139s
41. T. Ikeda, M. Nakano, Y. Yu, O. Tsutsumi, A. Kanazawa, Anisotropic bending and unbending behavior of azobenzene liquid-crystalline gels by light exposure. Adv. Mater. **15**(3), 201–205 (2003). https://doi.org/10.1002/adma.200390045
42. Z.Y. Wang, J. Zhang, X. Wu, M. Birau, G. Yu, H. Yu, Y. Qi, P. Desjardins, X. Meng, J.P. Gao et al., Near-infrared absorbing organic materials. Pure Appl. Chem. **76**(7–8), 1435–1443 (2004). https://doi.org/10.1351/pac200476071435
43. S.Q. Wang, D. Kaneko, M. Okajima, K. Yasaki, S. Tateyama, T. Kaneko, Hyperbranched polycoumarates with photofunctional multiple shape memory. Angew. Chem. Int. Ed. Engl. **52**(42), 11143–11148 (2013). https://doi.org/10.1002/anie.201305647
44. S. Masuo, S. Zhou, T. Kaneko, N. Takaya, Bacterial fermentation platform for producing artificial aromatic amines. Sci. Rep. **6**(1), 25764 (2016). https://doi.org/10.1038/srep25764
45. K. Takada, H. Shinagawa, Y. Morita, M.S. Grewal, K. Taya, A. Kumar, T. Kaneko, Syntheses of soluble biopolyimides using 4-aminophenylalanine. Chin. J. Polym. Sci. **38**(10), 1117–1123 (2020). https://doi.org/10.1007/s10118-020-2450-6
46. T. Kaneko, M.A. Ali, I. Captain, P. Perlin, T.J. Deming, Polypeptide gels incorporating the exotic functional aromatic amino acid 4-amino-l-phenylalanine. Polym. Chem. **9**(25), 3466–3472 (2018). https://doi.org/10.1039/C8PY00427G
47. T. Hirayama, A. Kumar, K. Takada, T. Kaneko, Morphology-controlled self-assembly and synthesis of biopolyimide particles from 4-amino-l-phenylalanine. ACS Omega **5**(5), 2187–2195 (2020). https://doi.org/10.1021/acsomega.9b03231
48. C.M. Hansen, A.L. Smith, Using Hansen solubility parameters to correlate solubility of C60 fullerene in organic solvents and in polymers. Carbon **42**(8), 1591–1597 (2004). https://doi.org/10.1016/j.carbon.2004.02.011
49. C.M. Hansen, 50 Years with solubility parameters—past and future. Prog. Org. Coat. **51**(1), 77–84 (2004). https://doi.org/10.1016/j.porgcoat.2004.05.004
50. C.M. Hansen, Aspects of solubility, surfaces and diffusion in polymers. Prog. Org. Coat. **51**(1), 55–66 (2004). https://doi.org/10.1016/j.porgcoat.2004.05.002

51. H. Yin, K. Takada, A. Kumar, T. Hirayama, T. Kaneko, Synthesis and solvent-controlled self-assembly of diketopiperazine-based polyamides from aspartame. RSC Adv. **11**(11), 5938–5946 (2021). https://doi.org/10.1039/d0ra10086b
52. G.M.J. Schmidt, 385. Topochemistry. Part III. The crystal chemistry of some trans-cinnamic acids. J. Chem. Soc. (Resumed) (0), 2014–2021 (1964). https://doi.org/10.1039/JR9640002014
53. F.D. Lewis, J.D. Oxman, Lewis acid enhancement of photochemical trans .fwdarw. cis isomerization of .alpha.,.beta.-unsaturated esters. J. Am. Chem. Soc. **103**(24), 7345–7347 (1981). https://doi.org/10.1021/ja00414a056
54. F.D. Lewis, S.L. Quillen, P.D. Hale, J.D. Oxman, Lewis acid catalysis of photochemical reactions. 7. Photodimerization and cross-cycloaddition of cinnamic esters. J. Am. Chem. Soc. **110**(4), 1261–1267 (2002). https://doi.org/10.1021/ja00212a039
55. P. Suvannasara, S. Tateyama, A. Miyasato, K. Matsumura, T. Shimoda, T. Ito, Y. Yamagata, T. Fujita, N. Takaya, T. Kaneko, Biobased polyimides from 4-aminocinnamic acid photodimer. Macromolecules **47**(5), 1586–1593 (2014). https://doi.org/10.1021/ma402499m
56. S. Dwivedi, T. Kaneko, Molecular design of soluble biopolyimide with high rigidity. Polymers (Basel) **10**(4) (2018). https://doi.org/10.3390/polym10040368
57. K. Takada, S. Karikome, S. Dwivedi, T. Kaneko, Fully bio-based aromatic polyimide using 4-aminocinnamic acid and mellophanic dianhydride as bio-derived monomers. ECS Trans. **88**(1), 99–105 (2018). https://doi.org/10.1149/08801.0099ecst
58. T. Noda, T. Iwasaki, K. Takada, T. Kaneko, Soluble biobased polyimides from diaminotruxinic acid with unique bending angles. Macromolecules **54**(22), 10271–10278 (2021). https://doi.org/10.1021/acs.macromol.1c01273
59. A. Isogai, T. Saito, H. Fukuzumi, TEMPO-oxidized cellulose nanofibers. Nanoscale **3**(1), 71–85 (2011). https://doi.org/10.1039/C0NR00583E
60. T. Saito, S. Kimura, Y. Nishiyama, A. Isogai, Cellulose nanofibers prepared by TEMPO-mediated oxidation of native cellulose. Biomacromol. **8**(8), 2485–2491 (2007). https://doi.org/10.1021/bm0703970
61. W. Chen, H. Yu, Y. Liu, Y. Hai, M. Zhang, P. Chen, Isolation and characterization of cellulose nanofibers from four plant cellulose fibers using a chemical-ultrasonic process. Cellulose **18**(2), 433–442 (2011). https://doi.org/10.1007/s10570-011-9497-z
62. M. Jonoobi, J. Harun, A.P. Mathew, M.Z.B. Hussein, K. Oksman, Preparation of cellulose nanofibers with hydrophobic surface characteristics. Cellulose **17**(2), 299–307 (2010). https://doi.org/10.1007/s10570-009-9387-9
63. A. Halperin, M. Kröger, F.M. Winnik, Poly(N-isopropylacrylamide) phase diagrams: fifty years of research. Angew. Chem. Int. Ed. **54**(51), 15342–15367 (2015). https://doi.org/10.1002/anie.201506663
64. H.G. Schild, Poly(N-isopropylacrylamide): experiment, theory and application. Prog. Polym. Sci. **17**(2), 163–249 (1992). https://doi.org/10.1016/0079-6700(92)90023-R
65. S. Kikuchi, Y. Chen, K. Fuchise, K. Takada, J. Kitakado, S.-I. Sato, T. Satoh, T. Kakuchi, Thermoresponsive properties of 3-, 4-, 6-, and 12-armed star-shaped poly[2-(dimethylamino)ethyl methacrylate]s prepared by core-first group transfer polymerization. Polym. Chem. **5**(16), 4701–4709 (2014). https://doi.org/10.1039/c4py00290c
66. L. Brannon-Peppas, N.A. Peppas, Dynamic and equilibrium swelling behaviour of pH-sensitive hydrogels containing 2-hydroxyethyl methacrylate. Biomaterials **11**(9), 635–644 (1990). https://doi.org/10.1016/0142-9612(90)90021-H
67. C. Zhu, L. Liu, Q. Yang, F. Lv, S. Wang, Water-soluble conjugated polymers for imaging, diagnosis, and therapy. Chem. Rev. **112**(8), 4687–4735 (2012). https://doi.org/10.1021/cr200263w
68. S. Dwivedi, A. Nag, S. Sakamoto, Y. Funahashi, T. Harimoto, K. Takada, T. Kaneko, High-temperature resistant water-soluble polymers derived from exotic amino acids. RSC Adv. **10**(62), 38069–38074 (2020). https://doi.org/10.1039/d0ra06620f
69. K. Takada, T. Noda, T. Kobayashi, T. Harimoto, M. Singh, T. Kaneko, Synthesis of pH-responsive polyimide hydrogel from bioderived amino acid. Polym. J. **53**, 1223–1230 (2021). https://doi.org/10.1038/s41428-021-00509-8
70. N. Yamano, A. Nakayama, N. Kawasaki, N. Yamamoto, S. Aiba, Mechanism and characterization of polyamide 4 degradation by Pseudomonas sp. J. Polym. Environ. **16**(2), 141–146 (2008). https://doi.org/10.1007/s10924-008-0090-y
71. T. Klement, J. Büchs, Itaconic acid—a biotechnological process in change. Biores. Technol. **135**, 422–431 (2013). https://doi.org/10.1016/j.biortech.2012.11.141
72. M.A. Ali, S. Tateyama, Y. Oka, D. Kaneko, M.K. Okajima, T. Kaneko, Syntheses of high-performance biopolyamides derived from itaconic acid and their environmental corrosion. Macromolecules **46**(10), 3719–3725 (2013). https://doi.org/10.1021/ma400395b
73. M.A. Ali, S. Tateyama, T. Kaneko, Syntheses of rigid-rod but degradable biopolyamides from itaconic acid with aromatic diamines. Polym. Degrad. Stab. **109**, 367–372 (2014). https://doi.org/10.1016/j.polymdegradstab.2014.05.031
74. J. Gao, B. Zhao, M.E. Itkis, E. Bekyarova, H. Hu, V. Kranak, A. Yu, R.C. Haddon, Chemical engineering of the single-walled carbon nanotube–nylon 6 interface. J. Am. Chem. Soc. **128**(23), 7492–7496 (2006). https://doi.org/10.1021/ja057484p

75. R. Jarrar, M.A. Mohsin, Y. Haik, Alteration of the mechanical and thermal properties of nylon 6/ nylon 6,6 blends by nanoclay. J. Appl. Polym. Sci. **124**(3), 1880–1890 (2012). https://doi.org/10.1002/app.35215
76. R. Greco, L. Nicolais, Glass transition temperature in nylons. Polymer **17**(12), 1049–1053 (1976). https://doi.org/10.1016/0032-3861(76)90005-7
77. H. Wang, K.W. Lee, T.-S. Chung, M. Jaffe, Rheology, morphology and properties of LCP/Nylon 66 composite fibers. Polym. Compos. **21**(1), 114–123 (2000). https://doi.org/10.1002/pc.10170
78. M.A. Ali, T. Kaneko, High-performance bionylons from itaconic and amino acids with pepsin degradability. Adv. Sustain. Syst. (2021). https://doi.org/10.1002/adsu.202100052
79. N. Saba, P.M. Tahir, M. Jawaid, A review on potentiality of nano filler/natural fiber filled polymer hybrid composites. Polymers **6**(8), 2247–2273 (2014)
80. V. Mittal, Polymer layered silicate nanocomposites: a review. Materials (Basel) **2**(3), 992–1057 (2009). https://doi.org/10.3390/ma2030992PMC
81. S. Pavlidou, C.D. Papaspyrides, A review on polymer-layered silicate nanocomposites. Prog. Polym. Sci. **33**(12), 1119–1198 (2008). https://doi.org/10.1016/j.progpolymsci.2008.07.008
82. S. Coiai, E. Passaglia, A. Pucci, G. Ruggeri, Nanocomposites based on thermoplastic polymers and functional nanofiller for sensor applications. Materials (Basel) **8**(6), 3377–3427 (2015). https://doi.org/10.3390/ma8063377PMC
83. M. Zhao, D. Yi, G. Camino, A. Frache, R. Yang, Interdigitated crystalline MMT–MCA in polyamide 6. RSC Adv. **7**(2), 861–869 (2017). https://doi.org/10.1039/C6RA26084E
84. S. Longo, M. Mauro, C. Daniel, M. Galimberti, G. Guerra, Clay exfoliation and polymer/clay aerogels by supercritical carbon dioxide. Front. Chem. **1**(28) (2013). https://doi.org/10.3389/fchem.2013.00028
85. T.D. Fornes, D.R. Paul, Structure and properties of nanocomposites based on nylon-11 and -12 compared with those based on nylon-6. Macromolecules **37**(20), 7698–7709 (2004). https://doi.org/10.1021/ma048757o
86. M.A. Ali, N. Tandon, T. Kaneko, Simultaneous hardening/ductilizing effects of cryogenic nanohybridization of biopolyamides with montmorillonites. ACS Omega **2**(12), 9103–9108 (2017). https://doi.org/10.1021/acsomega.7b01446.ws

Modification Techniques for Biomass-Based Plastics

Masayuki Yamaguchi

1 Viscoelastic Properties in the Molten State

Linear viscoelastic properties in the molten state are known to be important to comprehend the material characteristics of a polymer. Figure 1 shows the master curves of angular frequency ω dependence of oscillatory shear moduli, such as storage modulus G' and loss modulus G'', for some biomass-based polyesters such as PLA, PHB, and PBS [45, 53]. Furthermore, as a comparison, the oscillatory shear moduli of a commercially available isotactic polypropylene (PP) are also shown.

Although melt processing is applicable at low temperatures for PHB, the rheology measurement was not so easy at this temperature (180 °C) because of the chain-scission reaction. Therefore, the residence time dependence of the oscillatory moduli at each angular frequency was evaluated at first. Then, the G' and G'' curves as a function of the angular frequency were obtained considering the residence time in the rheometer.

The data in the figure are the values for the PHB sample with a minimized residence time (0 min). For PHB, the oscillatory moduli after the chain-scission reaction are superposed onto each other by the simple horizontal shift [45]. In the figure, a_D in the x-axis represents the horizontal shift factor by the degradation. As well known, M_w/M_n of most biomass-based polyesters is close to 2. Furthermore, a random chain-scission reaction leads to $M_w/M_n = 2$. As a result, the molecular weight distribution is unchanged even after the degradation. Therefore, the horizontal shift is applicable.

As seen in the figure, the rheological terminal zone is clearly detected for all samples in the low angular frequency region, i.e., $G' \propto \omega^2$ and $G'' \propto \omega$. Since G'' is proportional to the angular frequency, the zero-shear viscosity η_0 is obtained from the following equation.

$$\eta_0 = \lim_{\omega \to 0} \frac{G''}{\omega} \tag{1}$$

Beyond the critical molecular weight, which is approximately twice the average molecular weight between entanglement couplings M_e, η_0 is proportional to $M_w^{3.4}$. In fact, the following empirical relationships were reported; Eq. (2) for PLA at 180 °C [5] and Eq. (3) for PBS at 140 °C [7]

$$\ln \eta_0 = -32.835 + 3.4 \ln M_w \tag{2}$$

$$\ln \eta_0 = -32.595 + 3.31 \ln M_w \tag{3}$$

M. Yamaguchi (✉)
Research Center for Carbon Neutral, Japan Advanced Institute of Science and Technology, 1-1 Asahidai, Nomi 923-1292, Ishikawa, Japan
e-mail: m_yama@jaist.ac.jp

T. Kaneko (ed.), *Photo-switched Biodegradation of Bioplastics in Marine Environments*,
https://doi.org/10.1007/978-981-99-4354-8_4

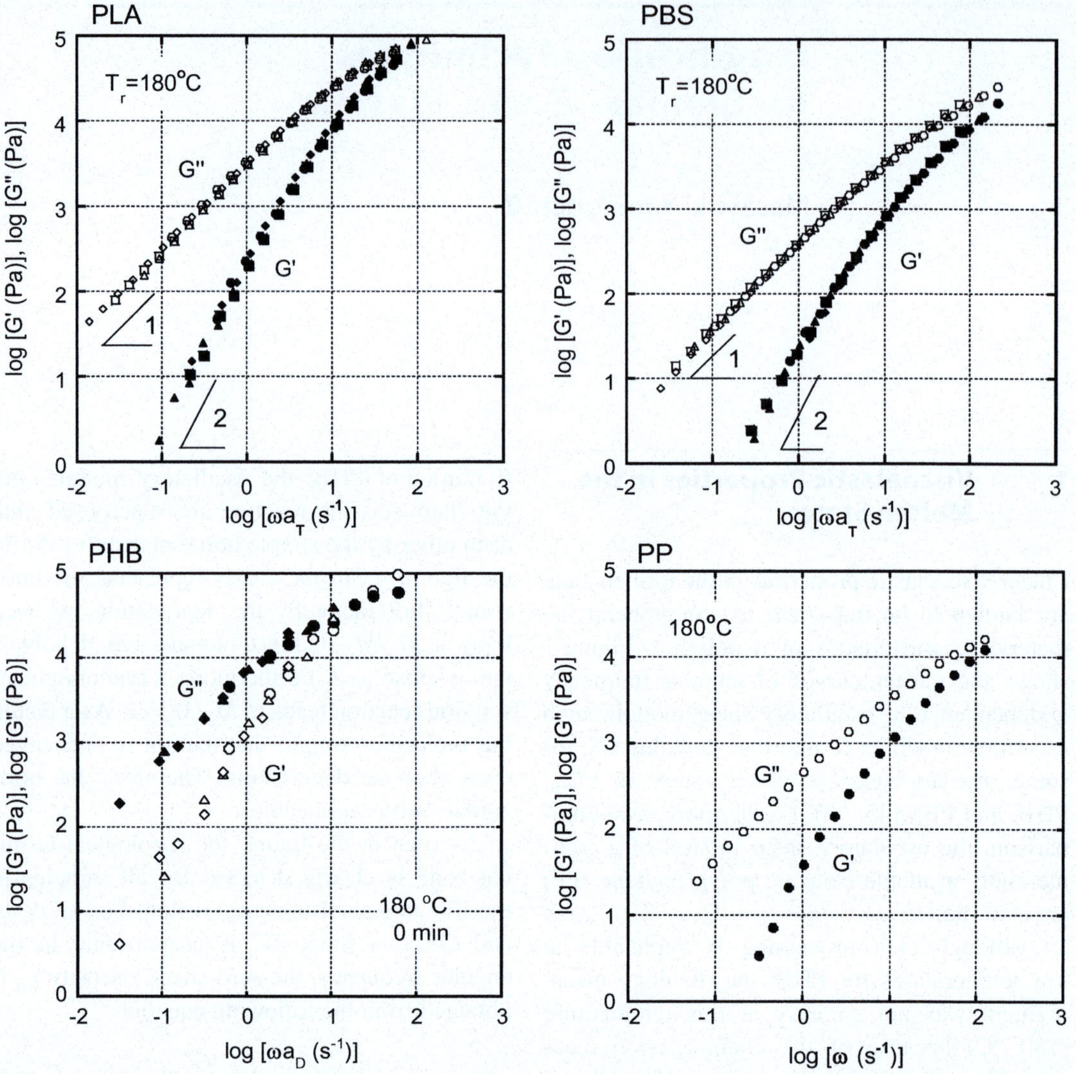

Fig. 1 Angular frequency ω dependence of oscillatory shear moduli such as storage modulus G' and loss modulus G'' at 180 °C for PLA, PBS, PHB, and PP at 180 °C [45, 53]

In the case of PLA and PBS, the master curves are shown by collecting the moduli at various temperatures. Following the Arrhenius-type Andrade equation (Eq. (4)), the flow activation energies ΔE_a are calculated from the shift factors a_T and found to be 67 kJ/mol for PLA and 43–45 kJ/mol for PBS [53], which correspond with the values reported previously [3, 7, 31]. According to Melik and Harrison [22], ΔE_a of PHB homopolymer is 37 kJ/mol, which decreases with the hexanoate content for PHBH, i.e., poly(3-hydroxybutyrate-*co*-3-hydroxyhexanoate). Janchai et al. [13] also reported a similar value for PHBH recently. For reference, ΔE_a of PP is 40 kJ/mol and that of HDPE is about 25 kJ/mol. Compared to them, the rheological properties of PLA are sensitive to the ambient temperature.

$$a_T = A \exp\left(-\frac{\Delta E_a}{RT}\right) \tag{4}$$

where A is a constant and R is the gas constant

The shear storage modulus G' is plotted as a function of loss modulus G'' in Fig. 2. As well known, G' represents the elastic nature while G'' shows the viscous one. Therefore, the figure represents the viscoelastic balance of each material. Although PP is known to show poor melt elasticity and thus needs to be modified its rheological properties for some processing operations such as foaming and blow-molding [50], the data of PP are located in the upper region. The results demonstrate that these biomass-based polyesters show poor melt elasticity. This is reasonable because their molecular weight distribution is quite narrow without long-chain branches. This is a typical problem not only for biomass-based polyesters but also for conventional polyesters produced from fossil resources. The modification techniques for this problem will be explained later.

Although there is no information in Fig. 1, the rubbery region appears in a much higher angular frequency region when the molecular weight is higher than the critical molecular weight. The rubbery plateau modulus G_N^0, i.e., storage modulus in the rubbery plateau region, is determined by M_e as follows:

$$G_N^0 = \frac{\rho RT}{M_e} = \nu_e k_B T \tag{5}$$

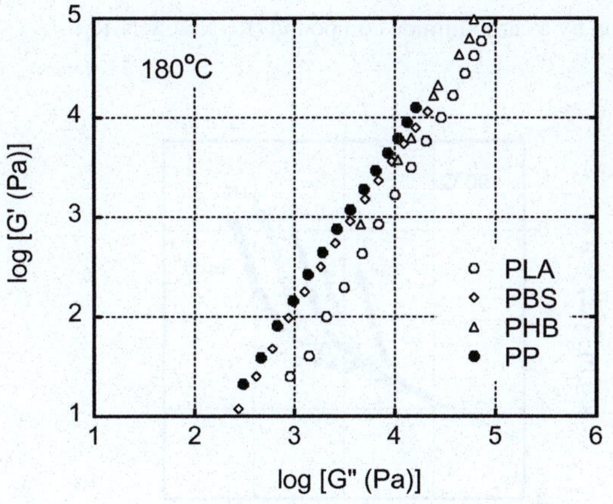

Fig. 2 Shear storage modulus G' plotted against loss modulus G'' at 180 °C for PLA, PBS, PHB, and PP

where ρ is the density, ν_e is the entanglement density, and k_B is the Boltzmann's constant.

The values of M_e were reported to be 8,000 for PBS [7] and 4,500–8,000 for PLA [5]. Yokohara and Yamaguchi [53] also found that M_e of PBS is higher than that of PLA. For PLA, it depends on the L-lactide content [4]. According to Liao et al. [20], M_e of PHBH decreases with 3-hydroxyhexanoate (HH) content from 11,600 (3.8 mol% of HH) to 9,400 (10.0 mol% of HH). These values are lower than those of polymethylmethacrylate (PMMA) ($M_e = 13{,}600$) and polystyrene (PS) ($M_e = 16{,}600$), and higher than those of bisphenol A polycarbonate (BPA-PC) ($M_e = 1660$) and poly(ethylene terephthalate) (PET) ($M_e = 1450$) [8, 9, 36]. As a comparison, M_e values of PP and HDPE are 6,850 and 1,040, respectively [8, 36]. The M_e value represents the flexibility of a polymer chain and is known to greatly affect mechanical properties in the solid state as well as viscoelastic properties in the molten state. For example, the critical stress of craze formation is believed to be proportional to $\nu_e^{1/2}$ [39]. Therefore, a polymer with high ν_e, i.e., low M_e, hardly shows brittle behavior in the glassy state. Moreover, the onset shear stress of shark-skin failure at the melt-extrusion process, σ_s, is known to be determined by G_N^0, as shown in Eq. (6) [1]. Therefore, a polymer melt with low M_e shows shark-skin failure only at a high shear rate.

$$\sigma_s = \frac{9}{4\pi} G_N^0 \frac{N_e}{\sqrt{N_0}} \tag{6}$$

where N_e is the number of entanglement couplings per chain and N_0 is the number of monomer unit per chain. To avoid the shark-skin failure, it is recommended to use a polymer having a broad molecular weight distribution [43]. Furthermore, processing aids such as fluoropolymer and boron nitride are preferably employed to avoid the shark-skin failure [15]. Recent topics to decrease the shear viscosity by the addition of a low-viscous immiscible polymer should be also noted [18, 33].

2 Modification of Rheological Properties and Processability

For melt processing, polyesters including biomass-based ones, have to be dried up enough to avoid the hydrolysis reaction, which is quite different from polyolefins. In the case of PLA, an equilibrium moisture content at 23 °C and 50% RH is around 2300 ppm, which is a similar level to that of PET [34]. For melt processing, the value must be lower than 250 ppm (has to be reduced more at high-temperature processing, e.g., 50 ppm for processing at 240 °C). Therefore, pellets have to be exposed to hot dry air, e.g., for a couple of hours at 90 °C [21]. Once the hydrolysis reaction occurs, the degradation grows much faster because of the catalyst role of acid compounds. The addition of a carbodiimide compound, which traps acid compounds, is a good method to reduce the degradation as shown in Fig. 3.

Besides the reduction of hydrolysis degradation, the enhancement of melt elasticity is required for biomass-based polyesters. One good method to evaluate the melt elasticity is the measurement of growth curves of uniaxial elongational viscosity [50]. Figure 4 shows the transient elongational viscosity η_E^+ at various elongational strain rates $\dot{\varepsilon}$ denoted by numerals [46]. The samples are a conventional PP without long-chain branches and a long-chain branched PP that was developed to provide pronounced melt elasticity. Basically, the growth curves of

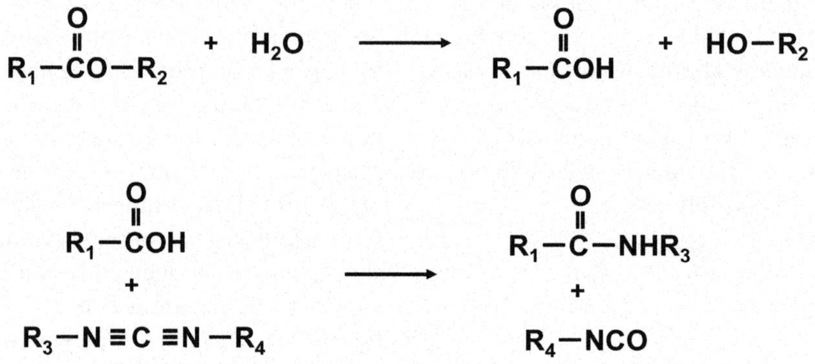

Fig. 3 Mechanism of the reduction of hydrolysis degradation by a carbodiimide compound (R_3-N≡C≡N-R_4)

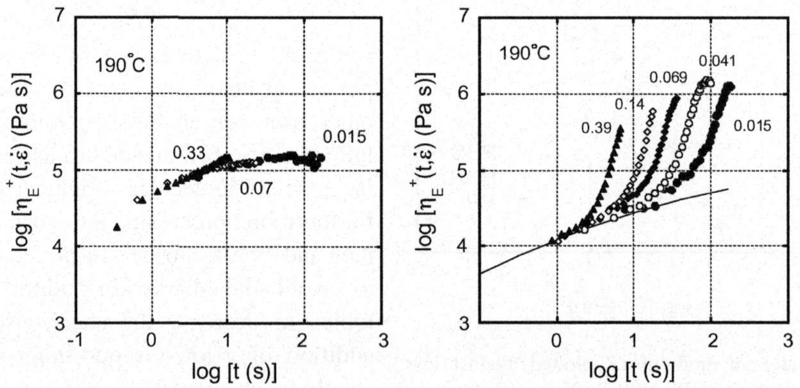

Fig. 4 Growth curves of uniaxial elongational viscosity η_E^+ at various elongational strain rates $\dot{\varepsilon}$ for (left) conventional PP and (right) long-chain branched PP at 190 °C. The numerals represent the elongational strain rates [46]

elongational viscosity are the same as those of the stress–strain curves, because time (x-axis) and viscosity (y-axis) are given by the strain divided by a strain rate and the stress divided by a strain rate, respectively.

As seen in the figure, elongational viscosity of a long-chain branched PP increases rapidly in a large strain (long time) region, which is called strain hardening, one of the most important elastic properties in the molten state. In contrast, all curves at different strain rates are superposed onto each other without any abrupt increase in the viscosity, i.e., no strain hardening, for a conventional PP. The strain hardening behavior, i.e., marked stress increase, is responsible for good processability at various processing operations [50], for example, the reduction of localized deformation and heat-sagging behavior at thermoforming and blow-molding, fine cell structure without coalescence of bubbles at foaming, the decrease in the neck-in and draw resonance at T-die film extrusion, and a stable bubble at a tubular-blown film. Therefore, a long-chain branched PP is preferred to be employed to improve the processability, especially for foaming and extrusion-coating. Low-density polyethylene with a lot of long-chain branches also shows good processability, with an intense strain hardening in transient elongational viscosity [42, 44].

The mechanism to improve the processability is exemplified in Fig. 5. When a polymer melt having a free surface is deformed, a large stress is concentrated in a thin part, leading to localized deformation. However, a polymer melt can be deformed homogeneously once it shows strain hardening, which results in a product with a homogeneous wall thickness. Bubble coalescence at foaming is reduced by the same mechanism.

The most conventional method to provide strain hardening is the introduction of long-chain branches. Because of chain stretching between branch points, strain hardening occurs [50]. For polyester, polyamide, and polycarbonate, a copolymer containing a glycidyl function is often employed. Because of the reaction between glycidyl function and the chain ends of these polymers, as shown in Fig. 6, branch structure is provided.

In industry, this reaction occurs in an extruder during processing, i.e., reactive extrusion [47]. Such compounds are commercially available, including Joncryl® from BASF, Arufon® from Toagosei, Lotader® from Arkema, and SAG® Fine-Blend Compatibilizer from Jiangsu.

Lamnawar and Maazouz [19] revealed that processing of a tubular-blown film becomes possible for PLA by the addition of a polymeric modifier containing glycidyl functions (Joncryl®) with a nucleating agent. Besides glycidyl compounds, isocyanate compounds, acid anhydrides, oxazoline compounds, and peroxide compounds could be employed for the same

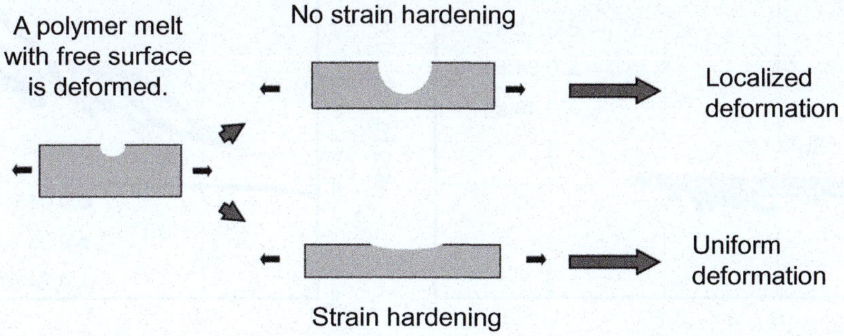

Fig. 5 Schematic illustration of a deformation mechanism of a polymer film with a free surface at melt processing; (top) a polymer melt without strain hardening results in a localized deformation and (bottom) a melt with strain hardening shows homogeneous deformation because a large deformation is prohibited by the rapid increase in the stress

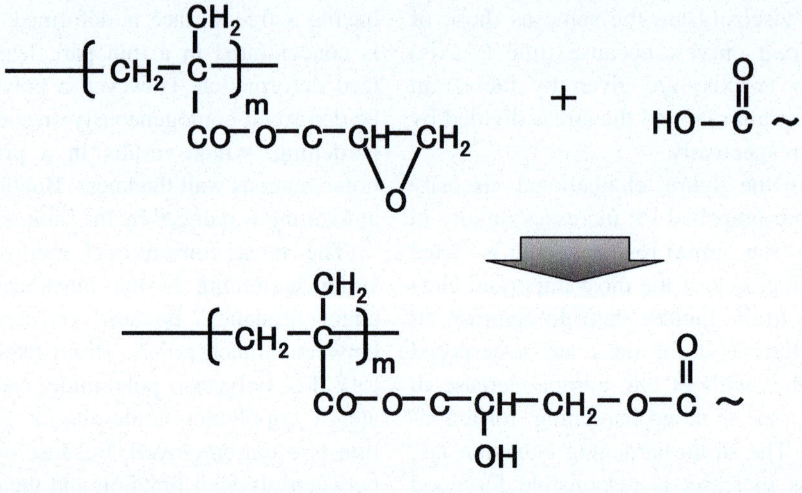

Fig. 6 Reaction mechanism between glycidyl function in a polymer and a chain end (acid) in a polyester

purpose [40]. In the case of bionylons, an oligomeric material containing maleic anhydride should be employed.

The addition of flexible nanofibers is also effective to provide strain hardening in elongational viscosity [23, 48, 49, 54]. Figure 7 exemplifies the growth curves of transient elongational viscosity for pure PLA and PLA containing 1 wt% of flexible poly(butylene terephthalate) (PBT) nanofibers [54]. In the figure, a growth curve of three times shear viscosity, $3\eta^+$, is denoted as a solid line, which is the growth curve of elongational viscosity at a low strain rate asymptote and calculated from oscillatory shear moduli, i.e., G' and G''. The deviation from the solid line is regarded as the degree

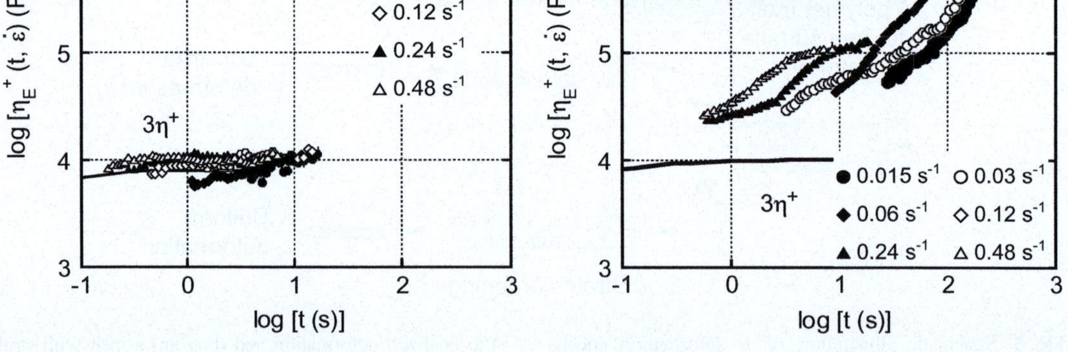

Fig. 7 Growth curves of uniaxial elongational viscosity η_E^+ at various elongational strain rates $\dot{\varepsilon}$ for (left) pure PLA and (right) PLA containing 1 wt% of PBT nanofibers at 180 °C. The numerals represent the elongational strain rates. An SEM image of the PBT nanofibers is also indicated in the right figure [54]

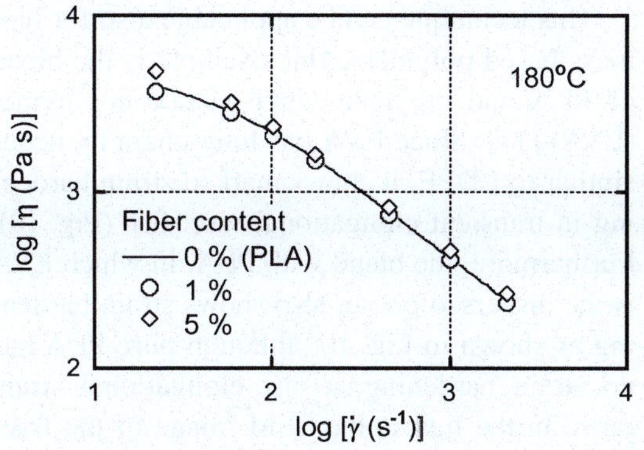

Fig. 8 Steady-state shear viscosity η as a function of shear rate $\dot{\gamma}$ at 180 °C for PLA, PLA containing 1 and 5 wt% of PBT nanofibers [54]

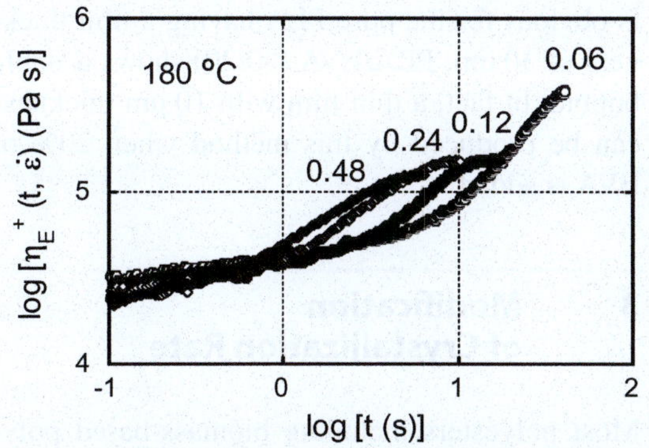

Fig. 9 Growth curves of uniaxial elongational viscosity η_E^+ at various elongational strain rates $\dot{\varepsilon}$ for PLA containing 1 wt% of PTFE nanofibers at 180 °C. The numerals represent the elongational strain rates [49]

of strain hardening. The measurement temperature is below the melting point of PBT (225 °C). Therefore, the PBT fibers are not melted during the measurements. The scanning electron microscope (SEM) image of PBT nanofibers after removal of PLA is shown in the right figure. The diameter is much smaller than 1 μm.

As seen in the transient elongational viscosity curves, strain hardening appears obviously by the nanofiber addition even though the amount of PBT fibers is only 1 wt%. It is also found from the viscoelastic measurements that these fibers have an interdigitated structure in a molten PLA. Therefore, the localized bending of PBT fibers and friction between fibers are inevitable during uniaxial flow, leading to the additional stress. This must be the origin of the strain hardening for this system [49].

Although this technique greatly affects the rheological response under elongational flow, the steady-state shear viscosity η is barely affected as shown in Fig. 8. It suggests that a torque level at extrusion and a flow length at injection molding are not affected greatly.

The PBT nanofibers are not commercially available, which was produced by hot-stretching of PLA/PBT blends beyond the melting point of PBT. At the temperature during hot-stretching, PBT shows lower viscosity. Therefore, it turns into a fibrous shape in the PLA matrix.

In industry, a specific powder of polytetrafluoroethylene (PTFE)-based polymer is available from Mitsubishi Chemical as Metablen™ A3000, which shows good dispersibility in a molten polymer including PLA [23]. When the powders are mixed with a molten PLA, the fibrous structure of PTFE is easily developed due to the weak cohesive strength of PTFE crystals [24]. Therefore, a time-consuming process, i.e., melt stretching beyond the melting point of PTFE, is not required. Of course, it provides strain hardening in elongational viscosity as shown in Fig. 9 [49] without the enhancement of shear viscosity.

A weak gel closed to a critical point of sol–gel transition, whose chain segments are miscible with the matrix polymer, is also a candidate for the modifier to enhance the melt elasticity [41]. Arakawa et al. [2] prepared a weak gel composed of poly(epichlorohydrin) (PECH), which is miscible with poly(3-hydroxybutyrate-co-3-hydroxyvalerate) (PHBV) in a molten state, and mixed with PHBV using a conventional internal mixer. They evaluated shear viscosity, extrusion swell ratio, and drawdown force defined as a force required to stretch a polymer melt uniaxially. The data are summarized in Table 1. The extrusion swell ratio is known to be enhanced with melt elasticity. Furthermore, the drawdown force has a close relation with elongational viscosity [48]. It is obvious that the addition of a weak gel of PECH, i.e., xPECH, greatly enhances the melt elasticity of PHBV.

Table 1 Extrudate swell ratio and drawdown force for PHBV and PHBV/xPECH [2]

Samples	Swell ratio	Drawdown force (mN)
PHBV	1.26	<10
PHBV/xPECH (98/2)	1.77	ca. 500

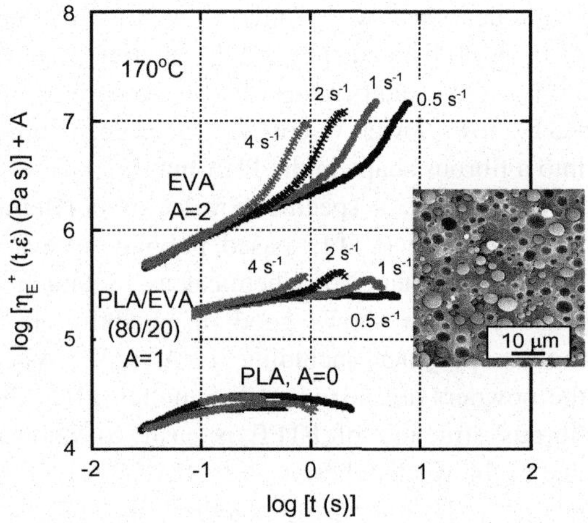

Fig. 10 Growth curves of uniaxial elongational viscosity η_E^+ at various elongational strain rates $\dot{\varepsilon}$ for (bottom) pure PLA, (top) pure EVA, and (center) PLA/EVA (80/20) at 170 °C. The numerals represent the strain rates. An SEM image of the fractured surface of PLA/EVA (80/20) is also shown

In contrast, shear viscosity is barely affected by xPECH [2]. For this technique, the mixing method should be considered because the dispersion has a higher viscosity than the matrix [50].

Blending with a long-chain branched polymer is another method to provide strain hardening in elongational viscosity, even when they are dispersed in a matrix [6]. In other words, dispersed droplets in an immiscible polymer blend has the capability to provide strain hardening in elongational viscosity. It seems to be strange because stress generation is greatly affected by a continuous phase, not a dispersed phase, in general. The detailed mechanism to show strain hardening was summarized in our previous papers [29, 51].

This technique can be applicable also for biomass-based polymers. One example is the blend of PLA and ethylene–vinyl acetate copolymer (EVA) [17]. Since EVA has long-chain branches similar to LDPE, it shows marked strain hardening in transient elongational viscosity (Fig. 10). Furthermore, the blend with PLA, in which EVA is the dispersed phase, also shows strain hardening as shown in Fig. 10, although pure PLA has no strain hardening at any elongational strain rates. In the figure, the SEM image of the fractured surface of the blend is shown. It is clear that EVA droplets are dispersed in PLA.

Because of the strain-hardening behavior, processability at a tubular-blown film is greatly improved [17]. Although the bubble instability is obvious for the pure PLA having a film thickness of 30 μm, PLA/EVA (80/20) shows a stable bubble. In fact, a thin film with 10 μm thickness can be produced by this method when 20% of EVA is added.

3 Modification of Crystallization Rate

Most polyesters including biomass-based polyesters show slow crystallization. Therefore, there is a strong demand in industry to enhance the crystallization rate. The crystallization rate is basically determined by the nucleating process of crystals and lateral growth rate. The former one is accelerated by the addition of an appropriate crystal nucleating agent. Furthermore, stereocomplex crystals of PLA having a high melting point also act as a nucleating agent for the rest of amorphous PLA [52]. Moreover, PHB crystals with a high melting point act as a nucleating agent for PHBV and PHBH [13]. For the lateral growth rate, which is usually evaluated by the growth rate of spherulite radius R_s, i.e., dR_s/dt, it is important to control the crystallization (cooling) temperature. In general, the lateral growth rate becomes a maximum at the following temperature $T_{c\text{-max}}$ [28],

$$T_{c-\max} = \frac{T_g + T_m^0}{2} \quad (7)$$

where T_g is the glass transition temperature and T_m^0 is the equilibrium melting point.

In the case of PLA, the lateral growth rate becomes a maximum at around 130 °C [10, 27]. However, even at the appropriate crystallization temperature, the growth rate of PLA is significantly lower than those of PE and PP. For a crystalline polymer with high T_g, the addition of a plasticizer is effective to enhance the growth rate. This phenomenon is pronounced especially at low crystallization temperatures, because T_g is shifted to a low temperature, leading to enhanced chain mobility. Considering the cost performance at processing operation, water should be employed as a cooling medium. Therefore, the plasticizer addition is favorable. However, the Flory–Huggins interaction parameter between a plasticizer and PLA should be positive. Otherwise, PLA crystals are dissolved in the plasticizer [10], leading to poor heat resistance due to low crystallinity.

Figure 11 shows the results of isothermal crystallization experiments for pure PLA and plasticized PLA samples [10, 27]. The lateral growth rate G_s (= dR_s/dt) is plotted against the crystallization temperature T_c. One plasticizer employed is a polyester with a molecular weight of 2000. Another one is an oligomeric polyethylene glycol capped by benzoate at chain ends. The latter decreases the melting point of PLA, suggesting that the interaction parameter with PLA is negative. Therefore, the crystallization rate is enhanced only in a restricted temperature region. In contrast, the addition of the polyester plasticizer greatly enhances the growth rate, especially in the low-temperature region.

Because of the enhanced lateral growth rate, the crystallization of PLA takes place even at the conventional cooling condition at compression molding. As a result, the heat resistance is greatly improved. Figure 12 shows the temperature dependence of the tensile storage modulus E' at 10 Hz [10]. All samples were prepared in the same cooling condition. In the case of pure PLA, the modulus greatly drops at around 60 °C, i.e., T_g. The modulus, then, increases at around 80 °C, which is attributed to the cold crystallization behavior. However, the heat-resistant temperature of this film is determined by T_g, because the material deforms downward by gravitational force beyond T_g. This is typical behavior of a quenched PLA sample. In contrast, PLA containing 20 wt% of the polyester plasticizer employed in Fig. 12 shows a gradual decrease in the modulus around T_g. Then, the modulus keeps almost a constant value until the melting point at 170 °C. The modulus curve of this film is similar to that of PP denoted in the

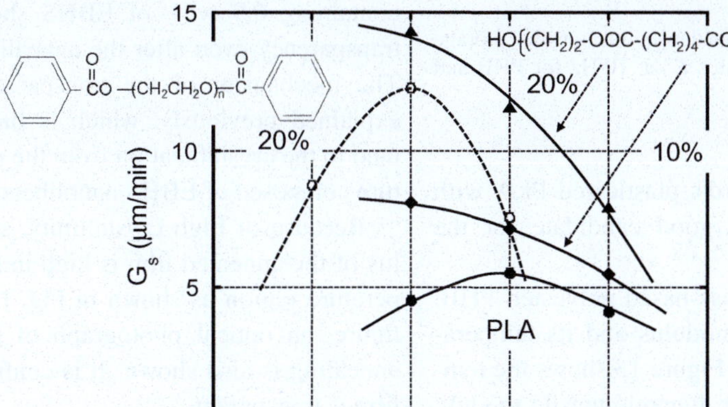

Fig. 11 Lateral growth rate G_s as a function of crystallization temperature T_c for PLA and plasticized PLA samples. The chemical structure of the plasticizers is denoted in the figure [10, 27]

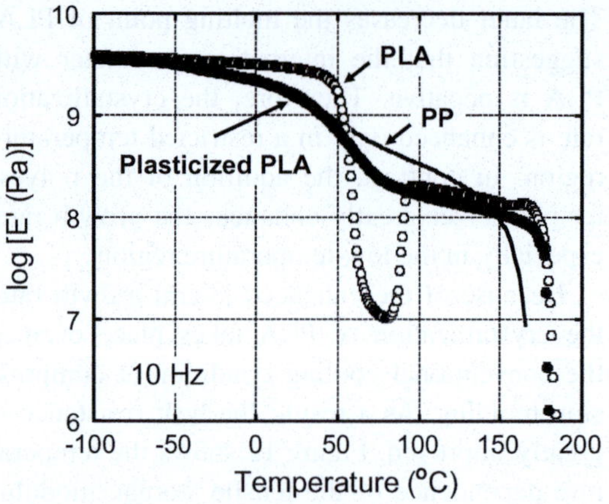

Fig. 12 Temperature dependence of tensile storage modulus E' at 10 Hz for pure PLA, plasticized PLA, and PP. All materials were cooled at the same condition [10]

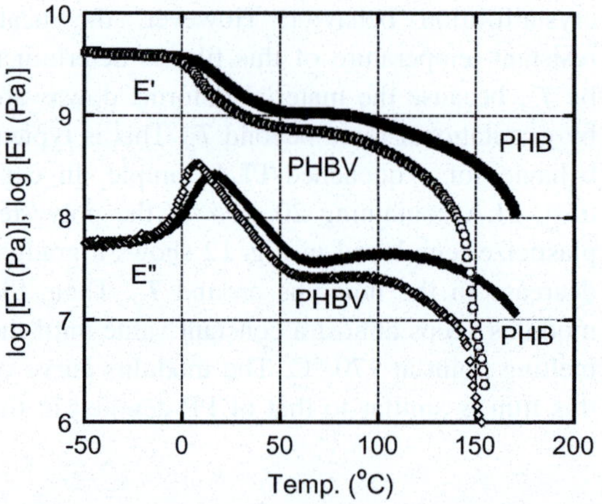

Fig. 13 Temperature dependence of tensile storage modulus E' and loss modulus E'' at 10 Hz for PHB and PHBV [2]

solid line. In other words, plasticized PLA with high crystallinity is a good candidate for the substitution of PP.

Another rigid biomass-based polyester, PHB, also shows a similar modulus and its temperature dependence to PP. Figure 13 shows the temperature dependence of dynamic tensile moduli, such as storage modulus E' and loss modulus E'', for PHB and PHBV having 12 mol% of 3-hydroxyvalerate (HV) at 10 Hz [2]. The peak temperature of the loss modulus E'' in the figure can be regarded as T_g. As compared with PHB, PHBV shows low E' values with a low melting point. Furthermore, the glass-to-rubber transition occurs at a low temperature. The E'' curve shows a sharp peak with a large area (relaxation strength) at T_g, indicating a large amount of amorphous region, i.e., low crystallinity. In the case of PHB, PHBH, and PHBV, mechanical properties at low temperature often needs to be required, because T_g is located near room temperature. A copolymer containing a large amount of HV or HH is thus recommended to be employed for such applications. Of course, the addition of a plasticizer may be another method to solve this problem.

When PLA is crystallized enough, the transparency is lost in general because of the light scattering. However, a unique method was proposed to prepare a crystallized PLA film with high transparency [32]. Figure 14 shows the haze values, i.e., the intensity of light scattering, for PLA films with 250 μm thickness, prepared by compression molding. The sample films contain various amounts of the nucleating agent denoted in the figure, i.e., N,N′-ethylenebis(12-hydroxystearamide) (EBHS). The films were quenched with/without post-processing annealing procedure at 100 °C for 2 h. As seen in the figure, when the EBHS content is low, the sample films are transparent, i.e., low haze values. This is reasonable because they are not crystallized enough. It should be noted that the film containing 0.9 wt% of EBHS shows excellent transparency even after the annealing procedure. The mechanism of the transparency was well explained previously, which is basically attributed to the crystallization from the network structure composed of EBHS nanofibers [32, 35].

Because of high crystallinity, storage modulus of the annealed film is high in the wide temperature region as shown in Fig. 15 [32]. In the figure, an optical photograph of the film after annealing is also shown. It is confirmed that the film is transparent.

A recent topic to enhance the crystallization rate is to use flow field. An applied flow field produces extended chain crystals when a strain rate is higher than the inverse of Rouse relaxation time [12–14, 37]. Since the Rouse relaxation time, i.e.,

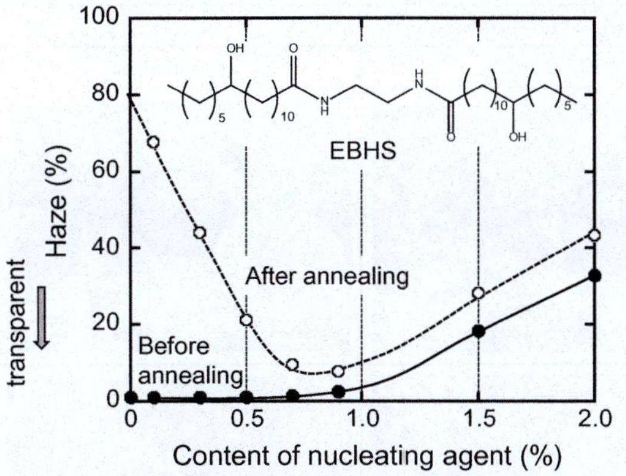

Fig. 14 Haze values of the PLA films containing various amounts of N,N′-ethylenebis(12-hydroxystearamide) (EBHS) before/after post-processing annealing procedure at 100 °C for 2 h. The film thickness is 250 μm [32]

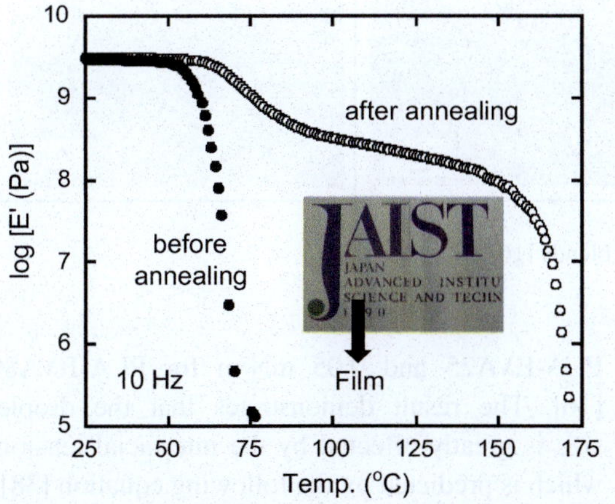

Fig. 15 Temperature dependence of tensile storage modulus E' at 10 Hz for the films containing 0.9 wt% of EBHS before/after post-processing annealing procedure at 100 °C for 2 h [32]

the characteristic time for chain contraction after a sudden deformation, is proportional to the square of molecular weight, adding a high-molecular-weight fraction is quite effective in enhancing the formation of extended chain crystals. Moreover, the performance of fibrous nucleating agents is enhanced in the flow field because more polymer chains easily attach to the surface of the nucleating agent because of their orientation to the flow field [11, 12, 30]. The oriented structure also contributes to high modulus [11, 26].

4 Modification of Mechanical Toughness

In general, the modification of mechanical toughness of biomass-based polyesters, especially PLA and PHB, is strongly desired for practical applications. Therefore, the rubber addition technique has been intensively studied. In particular, flexible polyesters produced from biomass resources, such as poly(butylene succinate) (PBS), poly(butylene succinate-*co*-butylene adipate) (PBSA), poly(ε-caprolactone) (PCL), poly(butylene adipate-*co*-butylene terephthalate) (PBAT), and poly(3-hydroxyalkanoate) with low/no crystallinity, are often employed, because they also show biodegradability. In many cases, however, a further addition of reactive compounds such as glycidyl compounds is required to improve their mechanical toughness because of poor compatibility. The mechanism is explained in Fig. 6 with commercially available compounds. Some blends containing both soft materials and reactive compounds have been already commercially available.

Hereafter, mechanisms of mechanical toughness improved by the rubber addition are explained using blend samples composed of PLA and EVA. The same concept is applicable to other blends composed of a rigid biomass-based plastic including bionylon and a flexible polymer.

PLA is known to be miscible with PMMA. Furthermore, PMMA is miscible with poly(vinyl acetate) (PVAc). Therefore, PLA shows good miscibility also with PVAc. In the case of EVA, relatively good compatibility was reported especially for those having a large amount of vinyl acetate (VAc) content. Figure 16 shows the SEM images of binary blends composed of PLA and EVA with various VAc contents [16]. The samples were prepared by melt mixing in an internal

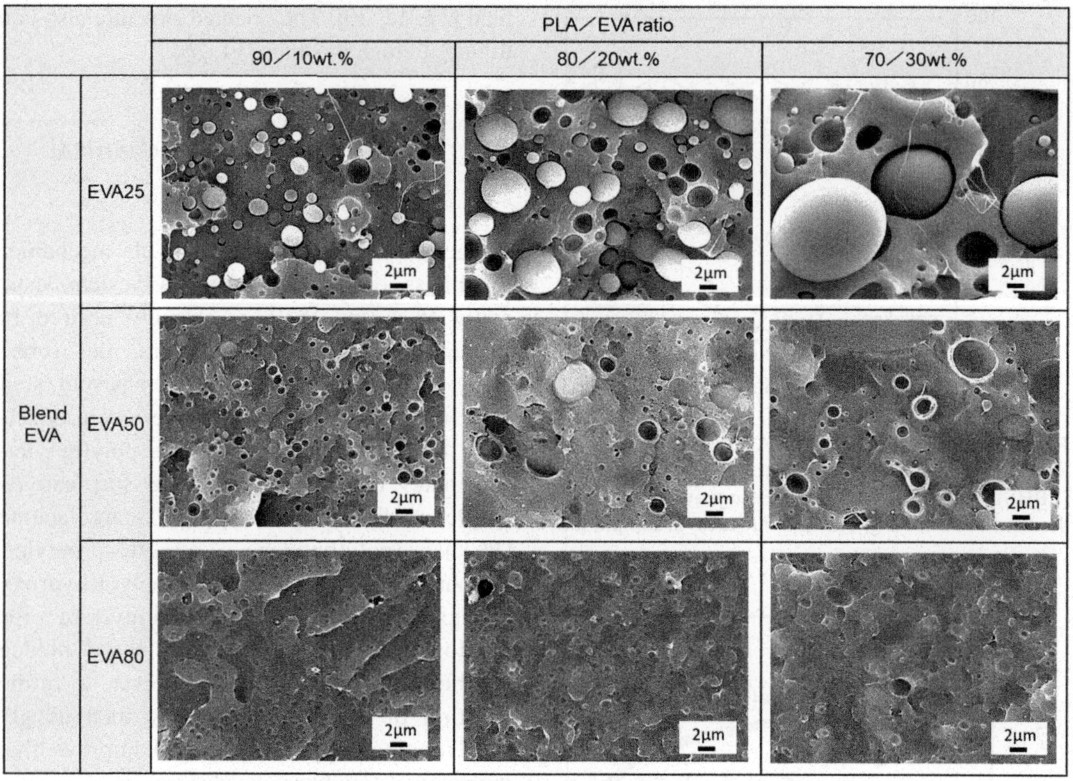

Fig. 16 SEM images of the fractured surface of PLA/EVA blends [16]

batch mixer, and then compressed into films with a quench condition. The numerals in the sample code represent the VAc content in EVA. In this experiment, the shear viscosity of PLA is higher than those of EVA samples at the mixing temperature. Furthermore, all EVA samples employed show almost the same shear viscosity. Therefore, the viscosity ratio of a matrix to dispersions, which affects the size of the dispersed phase in the blend [38], is not different greatly. The EVA contents are 10, 20, and 30 wt%. As seen in the figure, the droplet size increases with the EVA content. Moreover, the size decreases with increasing the VAc content in EVA.

The average size of dispersions is plotted in Fig. 17. It should be noted that the size of dispersions in the blends with EVA80 is quite small even in the blend with 30 wt% of EVA80 [16], suggesting good compatibility (Fig. 16).

The interfacial tensions, calculated from the surface tension, are found to be 2.4 mN/m for PLA-EVA25 and 0.05 mN/m for PLA-EVA80 [16]. The result demonstrates that the droplet size is greatly affected by the interfacial tension, which is predicted by the following equation [38]:

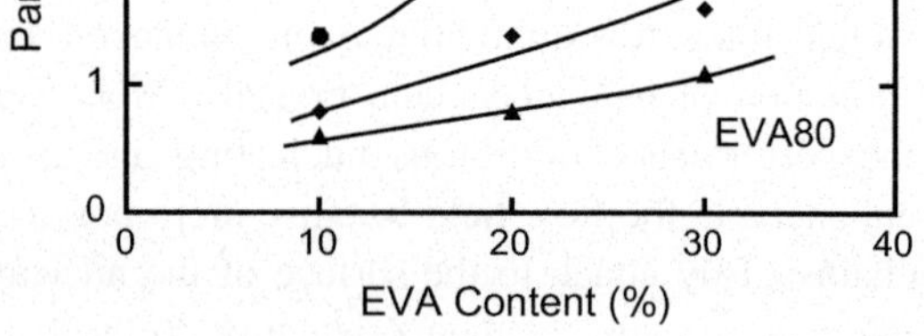

Fig. 17 Average particle size as a function of EVA contents in PLA/EVA blends [16]

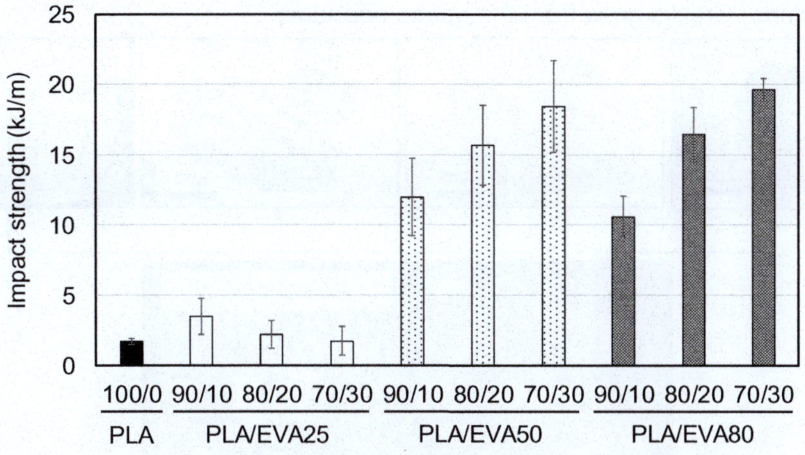

Fig. 18 Dart impact strength for the films of PLA, PLA/EVA25, PLA/EVA50, and PLA/EVA80 with various EVA contents [16]

$$R_d = 4\frac{\Gamma}{\sigma_m}\left(\frac{\eta_d}{\eta_m}\right)^{\pm 0.84} \tag{8}$$

where R_d is the radius of dispersion, Γ is the interfacial tension, σ_m is the shear stress exposed to the matrix, and η_d/η_m is the shear viscosity ratio of dispersion to the matrix.

The size of dispersion is reduced also by the addition of a compatibilizer including reactive compounds.

Figure 18 shows the impact strength of PLA and PLA/EVA films evaluated by the dart impact tests. Compared to PLA and PLA/EVA25, the other blends containing EVA50 and EVA80 exhibit pronounced toughness; i.e., EVA50 and EVA80 can improve the mechanical toughness of PLA to a great extent. Furthermore, it is interesting that the toughness for PLA/EVA25 decreases with the EVA25 content. The blend containing 30 wt% of EVA25 shows almost the same strength as the pure PLA. In contrast, the toughness increases with an increase in the EVA content for PLA/EVA50 and PLA/EVA80.

The morphology observation of the blends containing 20 wt% of EVA was carried out to confirm the fracture mechanism by a transmission electron microscope (TEM) after staining the sample films by RuO_4. As seen in the top images in Fig. 19, a crack propagates along the boundary of phases for PLA/EVA25. The picture indicates that the weak part in the blend is the phase boundary, which is ascribed to a thin interfacial thickness due to a low interfacial tension. The crack growth is the origin of the brittle fracture with a low impact strength. When numerous crazes appear from dispersed droplets, however, mechanical toughness is known to be improved. This mechanism is called multiple crazing, one of the important mechanisms to improve the toughness by the rubber addition, that is detected in acrylonitrile butadiene styrene terpolymer [25]. In this system, however, even for PLA/EVA80, the multiple crazing is not detected. Instead of multiple crazing, marked cavitation, i.e., void opening in dispersed droplets, is detected clearly in PLA/EVA80. The cavitation is another important mechanism for the toughness improvement by the rubber addition. The void formation due to a low cohesive strength in a rubbery particle reduces the actual strain applied to a material. As a result, the dilatational stress decreases greatly, which avoids brittle fracture.

Furthermore, the TEM image of PLA/EVA80 demonstrates that the PLA matrix is deformed greatly along with the deformation of rubber particles. The deformation of a matrix is the strong evidence for shear-yielding deformation.

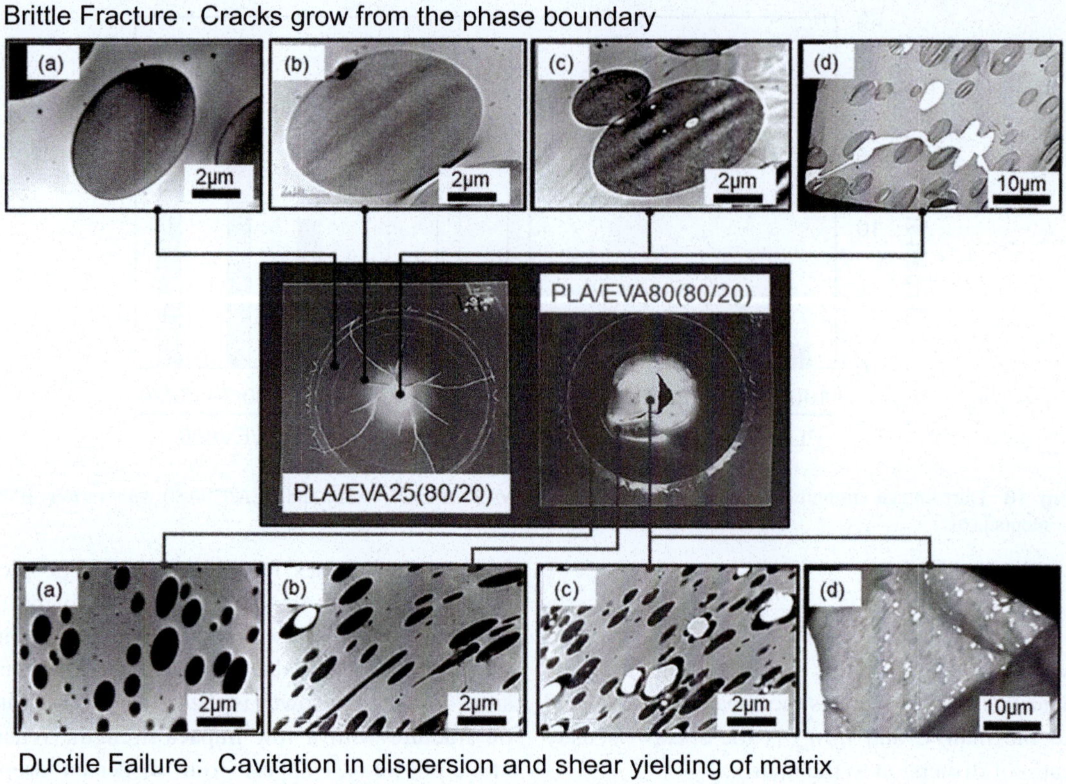

Fig. 19 Sample films after the dart impact tests for PLA/EVA25 (80/20) and PLA/EVA80 (80/20): (center) optical photographs, and TEM images of various positions for (top) PLA/EVA25 (80/20) and (bottom) PLA/EVA80 (80/20) [16]

The shear yielding is also an important mechanism to improve the mechanical toughness by the rubber addition. This phenomenon is explained using Fig. 20. In general, stress concentration occurs at the phase boundary, leading to the shear yielding of a matrix to some degree, which is denoted in the gray area in the right illustration. Then the stress concentration at the tip of the deformation area is enhanced more. When the ligament thickness, defined as the distance between the surfaces of neighboring rubber particles, is thick enough, the craze appears eventually once the concentrated stress is beyond the critical value of craze formation. The formation of crazes leads to brittle fracture. When the ligament thickness is thin, in contrast, a craze does not appear. Then, shear yielding area is connected to that of the neighbor one as shown in the right bottom illustration in Fig. 20. As a result, shear yielding occurs in all areas. In other words, the impact strength is improved when the ligament thickness is smaller than the critical value, irrespective of the rubber content. As demonstrated in Fig. 20, this blend system has the critical value for the ligament thickness, which is about 2 μm, to enhance the impact strength. The result indicates that the size of rubber particles should be small for the toughness improvement. Therefore, the addition of a compatibilizer is recommended when a soft material is added to a rigid plastic.

The abovementioned techniques for bioplastics should be applicable for bionylons described in Chap. 3. The research for preparing various plastics is ongoing.

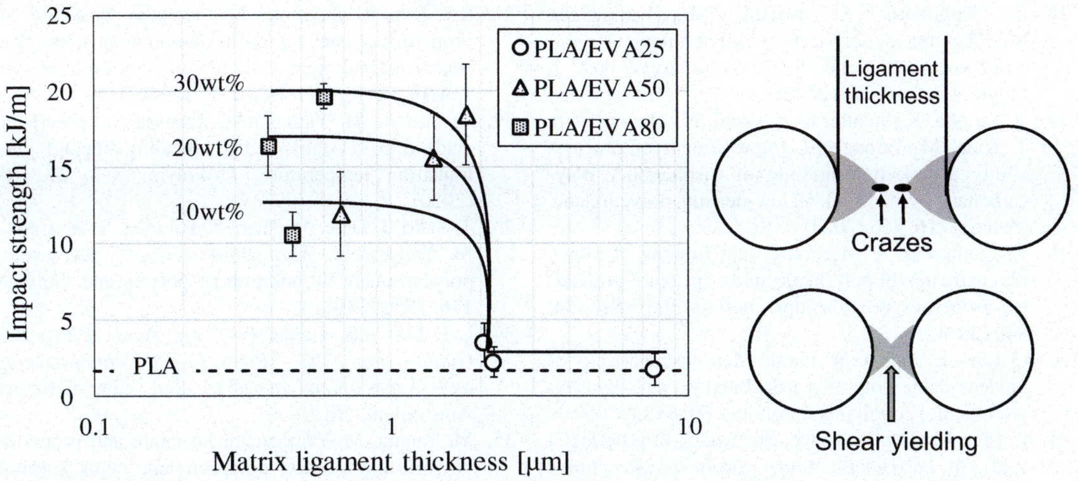

Fig. 20 Relationship between the matrix ligament thickness and the dart impact strength for the films of PLA/EVA blends [16]. The illustration in the right figure represents the mechanism of the toughness improvement considering the critical ligament thickness

Acknowledgements A part of this research was supported by Grant-in-Aid for Scientific Research (B) No. 22350102.

References

1. A. Allal, A. Lavernhe, B. Verges, G. Marin, Relationships between molecular structure and shark-skin defect for linear polymers. J. Non-Newtonian Fluid Mech. **134**, 127 (2006)
2. K. Arakawa, T. Yokohara, M. Yamaguchi, Enhancement of melt elasticity for poly(3-hydroxybutyrate-co-3-hydroxyvalerate) by addition of weak gel. J. Appl. Polym. Sci. **107**, 1320 (2008)
3. J.J. Cooper-White, M.E. Mackay, Rheological properties of poly(lactides). Effect of molecular weight and temperature on the viscoelasticity of poly(l-lactic acid). J. Polym. Sci., B, Polym. Phys. **37**, 1803 (1999)
4. J.R. Dorgan, J. Janzen, M.P. Clayton, S.B. Hait, D.M. Knauss, Melt rheology of variable L-content of poly(lactic acid). J. Rheol. **49**, 607 (2005)
5. J.R. Dorgan, Rheology of poly(lactic acid), in *Poly(Lactic Acid)—Synthesis, Structures, Properties, and Applications*, Chap. 10, ed. by R. Auras, L. Lim, S.E.M. Selke, H. Tsuji (Wiley, New York, 2010)
6. Y. Fujii, R. Nishikawa, P. Phulkerd, M. Yamaguchi, Modifying the rheological properties of polypropylene under elongational flow by adding polyethylene. J. Rheol. **63**, 11 (2019)
7. M. Garin, L. Tighzert, I. Vroman, S. Marinkovic, B. Estrine, The influence of molar mass on rheological and dilute solution properties of poly(butylene succinate). J. Appl. Polym. Sci. **131**, 40887 (2014)
8. W.W. Graessley, *Polymer Liquids & Networks: Dynamics and Rheology* (Garland Science, London, 2008)
9. H. Hasegawa, T. Kida, M. Yamaguchi, Novel transparent films composed of bisphenol-A polycarbonate and copolyester. Polymers **14**, 4146 (2022)
10. T. Huang, M. Miura, S. Nobukawa, M. Yamaguchi, Crystallization behavior and dynamic mechanical properties of poly(L-lactic acid) with poly(ethylene glycol) terminated by benzoate. J. Polym. Environ. **22**, 183 (2014)
11. S. Iwasaki, T. Inoue, Y. Takei, R. Nishikawa, M. Yamaguchi, Modulus enhancement of polypropylene by sorbitol nucleating agent in flow field. Polym. Cryst. **4**, e10170 (2021)
12. K. Janchai, T. Kida, T. Inoue, S. Iwasaki, M. Yamaguchi, Crystallization behavior of isotactic polypropylene containing a fibrous nucleating agent in a flow field. Polym. J. **54**, 367 (2022)
13. K. Janchai, T. Kida, M. Yamaguchi, T. Sunagawa, T. Okura, Optimum processing conditions for the maximum crystallization rate of poly(3-hydroxybutyrate-co-3-hydroxyhexanoate). Sci. Rep. **13**, 497 (2023)
14. T. Kida, T. Kimura, A. Eno, K. Janchai, M. Yamaguchi, Y. Otsuki, T. Kimura, T. Mizukawa, T. Murakami, K. Hato, T. Okawa, Effect of ultra-high-molecular-weight molecular chains on the morphology, crystallization, and mechanical properties of polypropylene. Polymers **13**, 4222 (2021)
15. R. Koopmans, J. den Doelder, J. Molenaar, *Polymer Melt Fracture* (CRC Press, Boca Raton, 2011)
16. D. Kugimoto, S. Kouda, M. Yamaguchi, Improvement of mechanical toughness of poly(lactic acid) by addition of ethylene-vinyl acetate copolymer. Polym. Test. **80**, 106021 (2019)

17. D. Kugimoto, S. Kouda, M. Yamaguchi, Modification of poly(lactic acid) rheological properties using ethylene-vinyl acetate copolymer. J. Polym. Environ. **29**, 121 (2021)
18. Y. Kuroda, K. Suzuki, G. Kikuchi, N. Moonprasith, T. Kida, M. Yamaguchi, Improvement of processability at injection-molding of bisphenol-A polycarbonate by addition of low-density polyethylene. Materials **16**, 866 (2023)
19. K. Lamnawar, A. Maazouz, Improvement of blown film extrusion of poly(lactic acid): structure-processing-properties relationships. Polym. Eng. Sci. **54**, 840 (2014)
20. Q. Liao, I. Noda, C.W. Frank, Melt viscoelasticity of biodegradable poly(3-hydroxybutyrate-*co*-3-hydroxyhexanoate) copolymers. Polymer **50**, 6139 (2009)
21. L. Lim, K. Cink, T. Vanyo, Processing of poly(lactic acid), in *Poly(Lactic Acid)—Synthesis, Structures, Properties, and Applications*, Chap. 14, ed. by R. Auras, L. Lim, S.E.M. Selke, H. Tsuji (Wiley, New York, 2010)
22. D. Melik, G.M. Harrison, Thermal degradation kinetics of poly(3-hydroxybutyrate-*co*-3-hydroxyhexanoate). J. Appl. Polym. Sci. **98**, 66 (2005)
23. B.M.A. Mohd Amran, K. Okamoto, M. Yamaguchi, A. Koshirai, T. Kasai, Rheological properties for polypropylene modified by polytetrafluoroethylene. J. Polym. Sci. Part B, Polym. Phys. Ed. **47**, 2008 (2009)
24. B.M.A. Mohd Amran, S. Nobukawa, M. Yamaguchi, Morphology development of polytetrafluoroethylene in a polypropylene melt. Pure Appl. Chem. **83**, 1817 (2011)
25. L. Morbitzer, D. Kranz, G. Humme, Structure and properties of abs polymers. X. Influence of particle size and graft structure on loss modulus temperature dependence and deformation behavior. J. Appl. Polym. Sci. **20**, 2691 (1976)
26. R. Nishikawa, M. Yamaguchi, Effect of carbon nanotube addition on structure and properties for extrudates of high-density polyethylene. J. Appl. Polym. Sci. **136**, 48010 (2019)
27. K. Okamoto, T. Ichikawa, T. Yokohara, M. Yamaguchi, Miscibility, mechanical and thermal properties of poly(lactic acid)/polyester-diol blends. Eur. Polym. J. **45**, 2304 (2009)
28. N. Okui, S. Umemoto, in *Polymer Crystallization*, Chap. 19, ed. by J.U. Sommer, G. Reiter (Springer, Berlin, 2003)
29. Y. Otsuki, Y. Fujii, H. Sasaki, P. Phulkerd, M. Yamaguchi, Experimental and numerical study on transient elongational viscosity for polypropylene/low-density polyethylene blends. Polym. J. **52**, 529 (2020)
30. P. Phulkerd, T. Nakabayashi, S. Iwasaki, M. Yamaguchi, Enhancement of drawdown force in polypropylene containing nucleating agent. J. Appl. Polym. Sci. **136**, 47295 (2018)
31. S.S. Ray, K. Okamoto, M. Okamoto, Structure and properties of nanocomposites based on poly(butylene succinate) and organically modified montmorillonite. J. Appl. Polym. Sci. **102**, 777 (2006)
32. T. Saitou, M. Yamaguchi, Transparent poly(lactic acid) film crystallized by annealing beyond glass transition temperature. J. Polym. Res. **27**, 104 (2020)
33. T. Sako, J. Date, M. Hagi, T. Hiraoka, S. Matsuoka, M. Yamaguchi, Anomalous viscosity decrease of polycarbonate by addition of polystyrene. Polymer **170**, 135 (2019)
34. L.T. Sin, B.S. Tueen, *Polylactic Acid: A Practical Guide for the Processing, Manufacturing, and Applications of PLA*, 2nd edn. (Elsevier, Amsterdam, 2019)
35. M. Tenma, M. Yamaguchi, Structure and properties of injection-molded polypropylene with sorbitol-based clarifier. Polym. Eng. Sci. **47**, 1441 (2007)
36. D.W. van Krevelen, K. te Nijenhuis, *Properties of Polymers*, 4th edn. (Elsevier, Amsterdam, 2008)
37. H.G.D. Vo, T. Kida, M. Yamaguchi, Role of shear flow on structure development during post-processing annealing for poly(lactic acid). Polymers **15**, 693 (2023)
38. S. Wu, Formation of dispersed phase in incompatible polymer blends: interfacial and rheological effects. Polym. Eng. Sci. **27**, 335 (1987)
39. S. Wu, Chain structure, phase morphology, and toughness relationships in polymers and blends. Polym. Eng. Sci. **30**, 753 (1990)
40. M. Xanthos, *Reactive Extrusion* (Hanser, Munich, 1992)
41. M. Yamaguchi, H. Miyata, Strain hardening behavior in elongational viscosity for binary blends of linear polymer and crosslinked polymer. Polym. J. **32**, 164 (2000)
42. M. Yamaguchi, M. Takahashi, Rheological properties of low density polyethylenes produced by tubular and vessel processes. Polymer **42**, 8663 (2001)
43. M. Yamaguchi, H. Miyata, V. Tan, C.G. Gogos, Relation between molecular structure and flow instability for ethylene-α-olefin copolymers. Polymer **43**, 5249 (2002)
44. M. Yamaguchi, D.B. Todd, C.G. Gogos, Rheological properties of LDPE processed by conventional processing machines. Adv. Polym. Technol. **22**, 179 (2003)
45. M. Yamaguchi, K. Arakawa, Effect of thermal degradation on rheological properties for poly(3-hydroxybutyrate). Eur. Polym. J. **42**, 1479 (2006)
46. M. Yamaguchi, M.H. Wagner, Impact of processing history on rheological properties for branched polypropylene. Polymer **47**, 3629 (2006)
47. M. Yamaguchi, T. Wakabayashi, Rheological properties and processability of chemically modified poly(ethylene terephthalate-*co*-ethylene isophthalate). Adv. Polym. Technol. **25**, 236 (2006)

48. M. Yamaguchi, K. Fukuda, T. Yokohara, B.M.A. Mohd Amran, S. Nobukawa, Modification of rheological properties under elongational flow by addition of polymeric fine fibers. Macromol. Mater. Eng. **297**, 654 (2012)
49. M. Yamaguchi, T. Yokohara, B.M.A. Mohd Amran, Effect of flexible fibers on rheological properties of poly(lactic acid) composites under elongational flow. J. Soc. Rheol., Japan **41**, 129 (2013)
50. M. Yamaguchi, Material strength in molten state for foam, in *Foam Extrusion: Principles and Practice*, 2nd edn., Chap. 4, ed. by S.T. Lee, C.B. Park (CRC Press, Boca Raton, 2014)
51. M. Yamaguchi, Y. Fujii, P. Phulkerd, Rheological properties of polyolefin blends containing long-chain branched polypropylene as dispersed phase. J. Soc. Rheol. Japan **48**, 109 (2020)
52. H. Yamane, K. Sakai, Effect of the addition of poly(D-lactic acid) on the thermal property of poly(L-lactic acid). Polymer **44**, 2569 (2003)
53. T. Yokohara, M. Yamaguchi, Structure and properties for biomass-based polyester blends of PLA and PBS. Eur. Polym. J. **44**, 677 (2008)
54. T. Yokohara, S. Nobukawa, M. Yamaguchi, Rheological properties of polymer composites with flexible fine fiber. J. Rheol. **55**, 1205 (2011)

Photoinduced Hydrophilicity and Antimicrobial Activity by Photocatalysis

Ken-ichi Katsumata, Shingo Machida, Kazuya Nakata and Makoto Ogawa

1 Introduction

Bacteria are responsible for the decomposition of biodegradable plastics, and the conditions for promoting the decomposition are preferably in a moist environment, that is, hydrophilic conditions. In addition, if biodegradable plastic is used and decomposition by bacteria progresses, the strength of the plastic cannot be maintained. That is, it is required to impart a hydrophilic condition at the time of decomposition and a function having an antibacterial effect at the time of use. In this chapter, we focus on the photocatalytic technology that expresses the above conditions by light energy, and explain its photoinduced hydrophilicity and antibacterial effect, for the purpose of photo-controlling the degradation of bionylons introduced in Chap. 3.

2 Photocatalytic Activity

The photocatalyst is generally used as a general term for substances that exhibit catalytic action by receiving photoexcitation. However, photocatalysts are also used for systems in which reactions that are thermodynamically impossible in a broad sense are possible by photoexcitation of the catalyst. That is, the photocatalyst has an aspect that a normal catalyst does not have, which enables a reaction that cannot occur thermodynamically in the dark by using a highly reactive active species generated by photoexcitation.

The photocatalytic activity of titanium oxide (TiO_2) is explained by the energy structure (band structure) of the semiconductor. The energy structure of a semiconductor has a band structure consisting of a valence band and a conduction band, and when it receives light with energy higher than the band gap, holes (h^+), which are electron shells, are generated in the valence band. Electrons (e^-) are generated in the conduction band. The valence band is mainly formed by the 2p orbital of oxygen, and the conduction band is formed by the 3d orbital of titanium (Fig. 1).

K. Katsumata (✉) · S. Machida
Department of Materials Science and Technology, Faculty of Advanced Engineering, Tokyo University of Science, 6-3-1 Niijuku Katsushika-Ku, Tokyo 125-8585, Japan
e-mail: k.katsumata@rs.tus.ac.jp

S. Machida
e-mail: shingo.machida@rs.tus.ac.jp

K. Nakata
Division of Sciences for Biological System, Institute of Agriculture, Tokyo University of Agriculture and Technology, 2-24-16 Naka-Cho, Koganei 184-0012, Tokyo, Japan
e-mail: nakata@go.tuat.ac.jp

M. Ogawa
School of Energy Science and Engineering, Vidyasirimedhi Institute of Science and Technology (VISTEC), 555 Moo 1 Tumbol Payupnai, Amphoe Wangchan, Rayong 21210, Thailand

T. Kaneko (ed.), *Photo-switched Biodegradation of Bioplastics in Marine Environments*,
https://doi.org/10.1007/978-981-99-4354-8_5

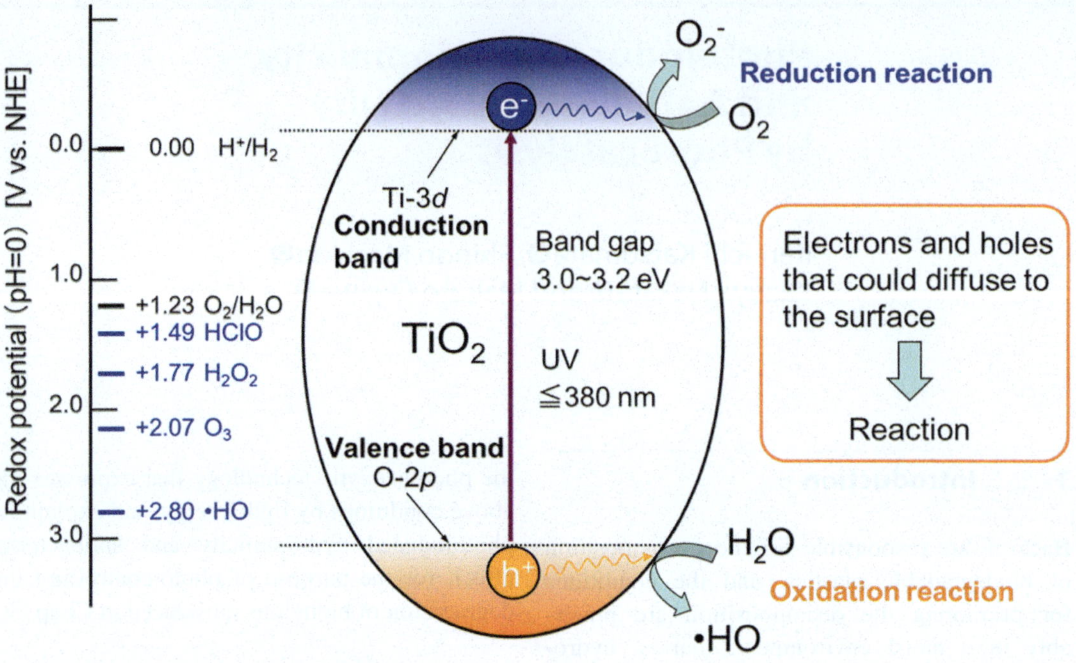

Fig. 1 Reaction mechanism of TiO_2 photocatalyst

For example, in the Honda-Fujishima effect [1], as a result of irradiating TiO_2 with ultraviolet light with a wavelength of 380 nm or less, which corresponds to the band gap, electrons, and holes are generated inside by electron excitation. Due to the Schottky barrier generated at the interface between TiO_2 and water, some of the electrons moved from the inside of TiO_2 to the counter electrode to generate hydrogen, while some of the holes moved to the surface of TiO_2 to oxidize water and generate oxygen. On the other hand, some of the electrons and holes generated by light irradiation in the atmosphere migrate to the TiO_2 surface without recombination due to the bending of the band generated by the adsorption of gas molecules on the surface. There is also a theory that these electrons and holes react with water and oxygen on the surface to generate various radicals such as –OH, O_2–, and –OH_2 (a part of the holes remains as surface supplement holes on the surface, and there is also a theory that atomic oxygen is generated), and these radicals decompose almost all organic substances. In the vicinity of the TiO_2 surface, the band bends due to the influence of adsorbed oxygen, and the electron–hole pairs generated by photoexcitation react with oxygen and water adsorbed on the TiO_2 surface to generate active oxygen species. do. These active oxygens have strong oxidizing power and can oxidatively decompose organic substances that compose dirt and bacteria [2–7]. This series of reactions is a catalytic reaction, and TiO_2 exhibits its effect durably without being consumed.

When the surface of TiO_2 formed into a thin film is irradiated with light and excited, the above-mentioned active oxygen is generated and diffused one after another, and the attached organic matter and adsorbed gas are oxidatively decomposed. By utilizing this reaction, it is possible to coat various materials with titanium oxide and impart functions such as antibacterial [5], stain decomposition [8], deodorization [9], and water washing [10].

On the other hand, as a reaction of a completely new titanium oxide photocatalyst, a phenomenon has been found in which the surface of TiO_2 becomes highly hydrophilic by light irradiation [11]. When the light irradiation is stopped, this hydrophilized surface gradually recovers to the original contact angle again. Regarding the mechanism of this phenomenon, it was initially thought that the organic substances adsorbed on the surface of TiO_2 were decomposed and removed by the photocatalytic reaction to produce a clean surface. That is, although the surface of a clean metal oxide is originally hydrophilic, mainly organic pollutants are adsorbed on the surface in the air, and as a result, the hydrophilicity is usually lowered. However, the idea is that the surface becomes clean due to the photocatalytic action of TiO_2, so that the original property of hydrophilicity appears. However, it cannot be explained by the idea based on such a simple photocatalytic reaction, and experimental facts suggesting that new surface changes are occurring have been reported one after another [12–15].

Such a super-hydrophilic surface is not only drip-proof and anti-fog, but also has a self-cleaning property that it is more compatible with water than oil and oil stains are easily removed with water. Currently, this phenomenon is being utilized to advance its application to films for automobile side mirrors, automobile body coatings, exterior tiles, and the like.

3 Photoinduced Hydrophilicity of Titanium Dioxide (TiO_2) Thin Films

Generally, the surface energy of metal oxides is large, and the surface of clean TiO_2 is considered to be highly hydrophilic in nature. Therefore, it was initially thought that the photoinduced hydrophilicization phenomenon of the TiO_2 thin film changes to hydrophilicity by decomposing pollutants adsorbed on the surface by a photocatalytic reaction and appearing a clean surface. However, in comparison with strontium titanate ($SrTiO_3$), which has almost the same bandgap and redox potential as TiO_2, TiO_2 shows oxidative decomposition activity and hydrophilization by light irradiation, whereas $SrTiO_3$ only exhibits oxidative decomposition activity [12]. It is known that it is not Sr but Ti–O that appears on the surface of $SrTiO_3$ fired in air [16], and it is considered that the surface composition of $SrTiO_3$ and TiO_2 is the same. In addition, even if TiO_2 is washed with strong alkali to completely remove the surface organic matter, the contact angle can only be increased to 10–20 degrees at most, and conversely, when light is irradiated, even if the surface organic matter remains, it is super. It has been shown to be hydrophilic [13]. From these facts, it is considered that the photoinduced super-hydrophilicity of TiO_2 is a phenomenon peculiar to this material.

With the aim of increasing the sensitivity of the photo-induced hydrophilization phenomenon of TiO_2, in order to make it hydrophilic even with weak ultraviolet light, there are reports on the design and synthesis of materials in which the charge separation efficiency is promoted and hydrophilicity is expressed under weak light by combining not only TiO_2 but also tungsten oxide (WO_3), whose band gap is narrow and both valence band and conduction band are located lower than TiO_2 [17, 18]. Also, by using a brookite-type TiO_2 sol that has become industrially synthesizable, studies on increasing the sensitivity with a brookite-type TiO_2 polycrystalline thin film that tends to create oxygen defects [19], repeated light irradiation [20], is being conducted. In addition, studies are being conducted to increase the sensitivity by multiplying and reflecting light in the film by laminating TiO_2, silica (SiO_2), and aluminum [21]. More recently, a small amount of nitrogen doping of TiO_2 has made it possible to obtain absorption in the visible region, and by optimizing the conditions, the hydrophilization reaction can proceed only with visible light [22–24]. However, this doping inevitably introduces strain into the structure, and as a result, although it is possible to impart a reaction with visible light, sensitivity to weak ultraviolet rays cannot be obtained.

The surface of TiO_2 that has become highly hydrophilic by light irradiation gradually returns to its original relatively hydrophobic state when stored in the dark. It is known that this hydrophobization rate is accelerated by various external perturbations. For example, when the temperature atmosphere for storage is changed, hydrophobization is promoted as the temperature rises. In addition, it is thought that surface stains will be removed and hydrophilicity will be improved by immersing the surface of titanium oxide, which has been made highly hydrophilic by light irradiation, in ultrapure water and applying ultrasonic waves, but in reality, the contact angle It has been reported that the temperature rises to about 10 degrees immediately [14]. Furthermore, when left in a vacuum, it becomes hydrophobic faster than in a normal pressure atmosphere.

From there, it obtained a moderately hydrophilic state with a contact angle of 10–20° at most by oxidative decomposition removal of contaminants adhering to the surface of TiO_2, and a contact angle higher than that. It is considered that the small, highly hydrophilized state is achieved by creating a metastable state with large surface energy by light irradiation [25].

However, it has been reported that the development of photoinduced super-hydrophilicity on the surface of TiO_2 is caused by the conventional photocatalytic reaction by the oxidative decomposition reaction of surface adsorption stains. Henderson et al. [26] reported the conversion of hydrophilicity and water repellency to the chemical adsorption of organic matter in high vacuum and the removal of organic matter by irradiation with ultraviolet rays. It is shown that when it is adsorbed, it becomes hydrophobic, and when it is irradiated with ultraviolet rays, it returns to hydrophilicity again. In addition, Shultz et al. [27] reported that oxidative decomposition of hydrocarbon-based films caused hydrophilicity by irradiation with ultraviolet rays in methanol vapor. However, these reports only indicate that the contact angle decreases with light irradiation, and there is no discussion that a highly hydrophilic state can be obtained.

On the other hand, Wang et al. [11] irradiate the surface of a single crystal of TiO_2 (rutile (110)) with ultraviolet rays, and the surface is surfaced with a friction force microscope (FFM). As a result of observation with FFM, the size of the super-hydrophilic surface after UV irradiation is about 20×50 nm in the [001] direction (a specific direction in which oxygen called unstable bridging oxygen is present). It was clarified that a hydrophilic-hydrophobic domain structure was generated. They also analyzed the surface state of the TiO_2 polycrystalline thin film (anatase) by X-ray photoelectron spectroscopy (XPS) and Fourier transform infrared spectroscopy (FT-IR). It has been reported that an increase in peaks suggesting an increase in water is observed on the surface of the TiO_2 thin film irradiated with light [28, 29]. Sato et al. [30] developed a special infrared absorption analyzer called Surface-enhanced infrared absorption spectroscopy (SEIRAS), which evaluated the surface of TiO_2 before and after light irradiation. It was demonstrated that water is photoadsorbed on the TiO_2 surface. In addition, Nosaka et al. [31–34] stated that photocatalytic action is related to the structure and characteristics of water adsorbed on the surface of TiO_2, and used nuclear magnetic resonance (NMR) for solids to determine the characteristics of adsorbed water. We are investigating and report that the structural water or surface-bonded hydroxyl layer on the surface of TiO_2 is destroyed by irradiation with ultraviolet rays. Furthermore, Uozaki et al. [35] have described sum frequency generation (SFG) spectroscopy, which is a second-order nonlinear vibration spectroscopy that is extremely sensitive to the interface, and quartz crystal microbalance (QCM), and evaluated the structure and amount of adsorbed water on the TiO_2 surface before and after light irradiation, and showed that both the amount and orientation of the adsorbed water increased by light irradiation. As described above, research on the photoinduced hydrophilization process by the surface chemical approach on the surface of TiO_2 is being actively conducted.

In addition, Sakai et al. [25] measured the water contact angle after using a TiO_2 polycrystalline thin film as an electrode and irradiating it with light while polarizing it. It was shown that hydrophilization occurs at a positive potential. This suggests that the holes generated by light irradiation contribute to the hydrophilization of TiO_2. Moreover, it has been clarified that the photo-induced hydrophilization reaction of TiO_2 thin films is affected by in-plane stress [36]. When compressive stress acts in the membrane due to the difference in expansion coefficient between the substrate and the membrane, it becomes difficult to become hydrophilic, and conversely, when tensile stress acts, it becomes easy to become hydrophilic. It is considered that this is because the relaxation of the bond by the hole is hindered by the surface compressive stress.

Based on these facts, the mechanism of photo-induced hydrophilization of the surface of TiO_2 is currently considered to be that holes among the electrons and holes generated inside TiO_2 weaken Ti-O bonds and water is dissociatively adsorbed there, resulting in hydrophilicity [25] as shown in Fig. 2.

On the surface of rutile single crystal, it is considered that the two-dimensional capillary effect is the essence of making the surface of TiO_2 highly hydrophilic due to the hydrophilic-hydrophilic regional structure revealed by AFM. On the other hand, anatase does not have a crystal plane having bridging oxygen like rutile, and from this, the photoinduced hydrophilicity of anatase obtained by the sol–gel method in the polycrystalline thin film is compared between the process of hydrophilization and the process of lipophilicity. Therefore, it is considered that the random arrangement of nano-sized particles in various crystal orientations causes variations in the rate of hydrophilization among the particles, which may produce a two-dimensional capillary effect [37, 38]. However, it has not yet been clarified why the rutile single crystal has a domain structure by light irradiation, and what is the relationship between the hydrophilization rate and the particle size in the anatase polycrystal. In the process of hydrophilicization of TiO_2, a hydrophilic lipophilic state (amphipathic state) is obtained in which the contact angle becomes 0 degrees with respect to both water and oil. It is understood that this is because there is a stage where two-dimensional capillaries are formed in both the hydrophilic part and the hydrophobic part in the process of forming the capillary of the hydrophilic part by light irradiation.

4 Photoinduced Hydrophilicity of Photocatalysts

There are not many reports on the photoinduced hydrophilicity of photocatalysts other than titanium oxide that have been studied in detail. Miyauchi et al. [39] investigated the relationship between the decomposition of methylene

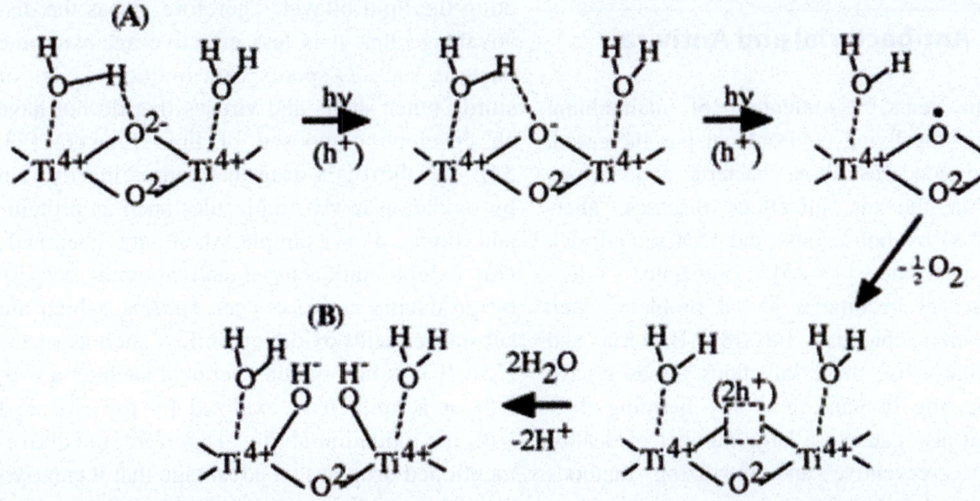

Fig. 2 Proposed mechanism for the photoinduced hydrophilic conversion of the TiO_2 surface [25]

blue and photoinduced hydrophilicity in 12 types of metal oxide thin films. Among them, it is reported that TiO_2 and ZnO exhibit methylene blue decomposition and photoinduced superhydrophilicity. On the other hand, SnO_2 and WO_3 are hydrophilic with almost no decomposition of methylene blue, but high super-hydrophilicity with a contact angle of 5 degrees or less has not been obtained. It has also been reported that $SrTiO_3$ exhibits methylene blue decomposition but does not undergo photoinduced hydrophilization. Regarding $SrTiO_3$, studies using single crystals have shown that the (100) and (111) planes have a faster hydrophilization rate under UV irradiation than the (110) planes, and are thought to be due to the unstable O atoms on the surface [40]

It has been reported that the $NaNbO_3$ polycrystalline thin film, which has the same perovskite-type structure as $SrTiO_3$, exhibits photoinduced super-hydrophilicity even though the decomposition activity of methylene blue is low when irradiated with ultraviolet rays (UV-A, UV-B) [41]. Furthermore, it is known that this highly hydrophilic surface lasts for about a week [42]. This result shows that it is different from the photo-induced hydrophilization mechanism of TiO_2. The mechanism of photoinduced hydrophilization of photocatalytic thin films remains unclear, and further research is desired.

5 Antibacterial and Antiviral

In recent years, as attention of maintaining hygiene in the living environment has increased, there are concerns about bacteria and viruses floating in the air, infectious diseases, allergies caused by house dust and VOCs, and sick house syndrome [43–45]. Especially, infectious diseases become a global problem when they become epidemic [46–48]. Bacteria and viruses that cause these infectious diseases have been infecting humans, not only harming their health but also causing a large number of deaths. Therefore, preventive and sterilizing methods for these infections have been required for a long time ago [49, 50].

One of the methods that have been used for a long time to inactivate bacteria and viruses is an antibacterial and antiviral agent, respectively. There are various types of antibacterial and antiviral agents that are developed and used. Here, we have classified them into three main categories according to their inactivation mechanism. The first is a drug that inhibits the activity of bacteria or viruses, such as antibiotics and antivirus drugs for bacteria and viruses, respectively. For example, penicillin specifically binds to an enzyme that synthesizes peptidoglycan, which is a major component of the bacterial cell wall, and inhibits its activity to suppress its growth [51, 52]. This is a highly specific reaction and has the advantage of being almost harmless to the human body. However, due to its high specificity, there is a drawback that drug-resistant bacteria, due to gene mutation, are likely to generate. The second is a drug involved in the destruction of bacteria and viruses and the denaturation of molecules in their body. For example, ethanol causes inactivation of bacteria and viruses by destroying cell membranes and denaturing proteins. This drug has the advantage of high immediate effect because it begins to react with the cell membranes and proteins of the target bacteria and viruses immediately after being added. However, alcohols such as ethanol are targeted at the lipid bilayer itself or proteins inside bacteria and viruses after permeating the lipid bilayer. Therefore, it has the disadvantage that it is less effective against some bacteria such as spores with multiple layers of sturdy outer shells and viruses that do not have an envelope composed of lipid bilayers [53, 54]. The third is a drug that causes inactivation by oxidation in vivo molecules such as proteins and lipids. For example, there are chemicals that exhibit antibacterial and antiviral activity by producing active oxygen species, which are substances with oxidation ability, such as ozone [55]. If it is an organic chemical such as a protein or a lipid, it is oxidized by the chemical with the oxidation ability. Therefore, the above-mentioned drug has an advantage that it can oxidize almost all bacteria and viruses composed of organic substances and has antibacterial and antiviral activities against a very wide variety of bacteria and viruses. In addition, since many bacteria and viruses are composed of organic chemicals, almost all bacteria and viruses are non-specifically oxidized and inactivated, so there is an advantage that resistant bacteria and resistant viruses are less likely to occur. On the other hand, drugs that utilize oxidation have the disadvantage that they may be harmful to humans because they cause an oxidative reaction not only with bacterial and viral proteins but also with proteins in the human body. As mentioned above, antibacterial and antiviral agents have different advantages and disadvantages. These antibacterial and antiviral agents have been used properly according to the type of target bacteria and viruses, the surrounding environment and the situation. Research on antibacterial and antiviral agents that can be applied to specific environments and situations is still proceeding.

This chapter describes photocatalysts as antibacterial and antiviral agents. A photocatalyst is a drug that can generate holes and reactive oxygen species under light irradiation and inactivate various bacteria and viruses. Since this drug produces reactive oxygen species having the oxidizing ability, it is classified as a drug that inactivates bacteria and viruses by oxidation in the above classification, and is expected as an antibacterial and antiviral agent.

Since the Honda-Fujishima effect was discovered in 1967 and reported in Nature in 1972 [1], various fundamental types of research have been carried out and practical application of photocatalysts has been promoted. Currently, the most used photocatalyst is TiO_2, which is practically used in air purifiers, windows, wall tiles, etc. [56–58]. Products related to photocatalysts cover a wide range of fields such as housing, electrical appliances, vehicles, roads, agriculture, water treatment, daily necessities, and medical fields. It shows that research in the field of photocatalysts is being actively carried out [59].

There are two major functions of photocatalysts, "oxidative decomposition ability" and "super-hydrophilicity". The former oxidative decomposition ability is a function that a photocatalyst can decompose organic chemicals and finally convert them into carbon dioxide and water, so that air purifiers utilize this property to prevent harmful substances, bacteria and viruses in the air. The latter superhydrophilicity is a phenomenon in which the water contact angle becomes 5° or less and is applied to self-cleaning using a photocatalyst [11].

6 Antibacterial and Antiviral Activity with Photocatalyst

As described above, the photocatalyst induces oxidative and reductive reactions due to electrons, holes, and active oxygen species. Initially, research on photocatalysts mainly carried out hydrogen generation using the reduction reaction in the field of water splitting, but after that, organic chemical decomposition using oxidative reaction began to be carried out [60, 61]. In this decomposition of organic chemicals, holes, and reactive oxygen species oxidize it and finally decompose it into carbon dioxide and water. In particular, hydroxyl radicals have a large oxidative energy equivalent to 120 kcal/mol, whose energy is larger than the bonding energy (around 100 kcal/mol) of carbon–carbon, carbon–nitrogen, carbon-hydrogen, oxygen-hydrogen, nitrogen–hydrogen bondings, etc. which means these bonds can be easily cleavage. Therefore, since photocatalysts can decompose VOCs such as acetaldehyde and other pollutants, their application to air purification is currently being promoted.

In principle, photocatalysts can decompose various organic chemicals, but they also have the problem that complete decomposition takes a long time. Based on this background, the advantages and disadvantages of photocatalysts were reconsidered, and the application of TiO_2 to antibacterial treatment was carried out for the first time in 1985 [62]. In antibacterial activity using TiO_2, if the bacteria can be partially damaged even if not all the cells can be decomposed, the bacteria may be inactivated, which does not take much time than the complete decomposition. In fact, TiO_2 showed an effective inactivating effect on E. coli [63]. Cho, M et al. reported inactivation of E. coli with TiO_2 [64]. The survival rate of E. coli decreased under ultraviolet irradiation, which suggested TiO_2 inactivated E. coli. Subsequent studies have reported that it is also effective against other bacterial species and protists [5, 65–67]. The virus inactivation was first reported in 1995 [68]. Currently, applied research is being actively conducted with the aim of creating a hygienic environment using photocatalysts.

In parallel with the above-mentioned research on the application of TiO_2 to antibacterial activity, research on the antibacterial mechanism with TiO_2 was also conducted. Initially, it was proposed that the antibacterial mechanism with TiO_2 was caused by the inactivation of the coenzyme CoA [62], but now it is thought that cell membrane damage is the main reason [69, 70]. One of the mechanisms is proposed in Fig. 2 [4]. Firstly, the cell membrane is destroyed and then intracellular substances flow out to inactivate it (note that the presence of the peptidoglycan layer is not related to the inactivation in this paper). However, (1) the hypothesis of inactivation of coenzyme CoA has not been denied, (2) intracellular enzyme oxidation by the Fenton reaction has been newly proposed [71], (3) functional proteins in bacteria are also inactivated at the stage of membrane damage [72], and (4) the cell membrane damage hypothesis is not perfect because the possibility of bacterial inactivation before the membrane is damaged cannot be ruled out. From the above points, the whole picture of the antibacterial mechanism has not been clarified yet. In addition, as described in 3), it has been reported that the activity of the functional protein contained in the bacterium is lost, but whether the bacterium is inactivated only by the loss of the activity of the functional protein, etc. At present, the details have not been discussed. In any case, it is now generally accepted that the destruction of bacterial external structures such as the outer cell membrane and cell membrane by reactive oxygen species is the main factor of inactivation [67]. Holes with strong oxidizing ability are also generated by the photocatalytic reaction, but since the holes are generated on the surface of the photocatalyst, they react only with the bacteria in contact with them and immediately react with the surrounding water. It is thought that reactive oxygen species mainly contribute to the photocatalytic antibacterial activity than holes (Fig. 3).

On the other hand, in studies on virus inactivation, bacteriophage (the virus that infects bacteria, hereinafter referred to as phage) was often used as a substitute for the virus [65, 73, 74]. This is because bacteriophage Qβ (hereinafter referred to as Qβ phage) is used for antivirus performance evaluation in the JIS standard

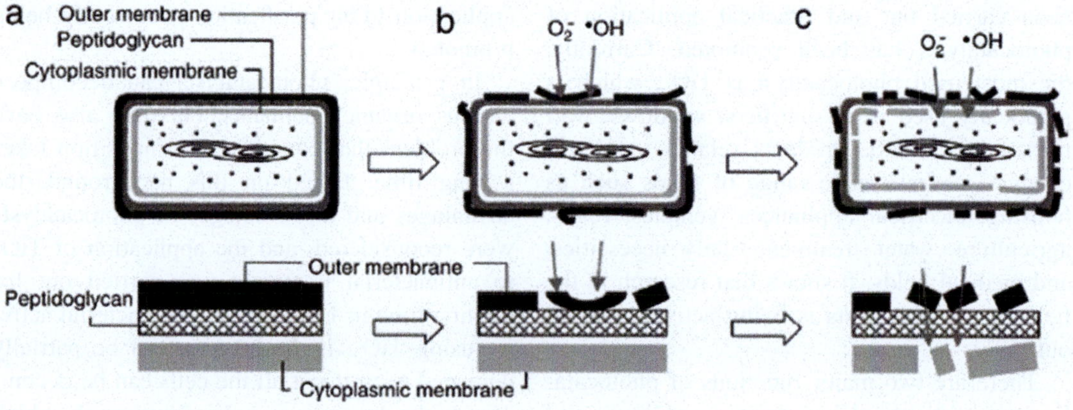

Fig. 3 Antibacterial mechanism with photocatalyst [4]

of ceramics, and many academic papers report phage as a substitute for virus [75–77]. Cho M. et al. reported inactivation of MS2 phage with TiO_2 [64]. Similar to E. coli, the survival rate of MS2 phage decreased under UV irradiation in the presence of TiO_2. However, there are few reports on the photocatalytic anti-phage mechanism. For example, it is said that the capsid protein forming the outer shell of the phage is damaged by the reactive oxygen species generated by the photocatalytic reaction [76, 78]. On the other hand, although it has been reported that the nucleic acid inside the phage is not damaged within the range where the infectious titer of the phage is reduced, there are few reports on the protein that works inside the phage [79], thus the study on the inside of the phage is insufficient. Currently, as in the case of bacteria, the destruction of the external structure of phage with reactive oxygen species is considered to be the main reason of its inactivation [70]. In addition, there are few reports on antibacterial and antiviral mechanisms using visible light-responsive photocatalysts, and research has not progressed much [80, 81].

There have been some reports of photocatalytic inactivation against viruses compared to phages [4, 60]. In addition, due to the global influenza pandemic in 2009, research on photocatalytic virus inactivation has been actively conducted [82, 83]. The mechanism of influenza virus inactivation with TiO_2 is not considered to have a large effect on nucleic acids, as reported in phage [83]. It has also been proposed that this is due to damage to the outer shell proteins of virus particles, specifically neuraminidase and hemagglutinin, by reactive oxygen species produced by photocatalytic reactions. Although the reaction with neuraminidase and hemagglutinin has been mentioned, the specific mechanism remains unclear. Basically, TiO_2 generates a large amount of reactive oxygen species together with holes by the photocatalysis, so that their external structures are non-specifically destroyed regardless of the type of virus, which is a widely accepted mechanism for the inactivation.

References

1. A. Fujishima, K. Honda, Electrochemical photolysis of water at a semiconductor electrode. Nature **238**, 37–38 (1972)
2. T. Kawai, T. Sakata, Conversion of carbohydrate into hydrogen fuel by a photocatalytic process. Nature **286**, 474–476 (1980)
3. J. Schwitzgebel, J.G. Ekerdt, H. Gerischer, A. Heller, Role of the oxygen molecule and of the photogenerated electron in TiO_2-photocatalyzed air oxidation reactions. J. Phys. Chem. **99**(15), 5633–5638 (1995)
4. K. Sunada, T. Watanabe, K. Hashimoto, Studies on photokilling of bacteria on TiO_2 thin film. J. Photochem. Photobiol. A: Chem. **156**(1–3), 227–233 (2003)
5. K. Sunada, Y. Kikuchi, K. Hashimoto, A. Fujishima, Bactericidal and detoxification effects of TiO_2 thin film photocatalysts. Environ. Sci. Technol. **32**(5), 726–728 (1998)
6. C.C. Trapalis, P. Keivanidis, G. Kordas, M. Zaharescu, M. Crisan, A. Szatvanyi, M. Gartner, $TiO_2(Fe^{3+})$ nanostructured thin films with antibacterial properties. Thin Solid Films **433**(1–2), 186–190 (2003)
7. A. Mills, S. LeHunte, An overview of semiconductor photocatalysis. J. Photochem. Photobiol. A: Chem. **108**(1), 1–35 (1997)
8. T. Minabe, D.A. Tryk, P.S. Sawunyama, Y. Kikuchi, K. Hashimoto, A. Fujishima, TiO_2-mediated photodegradation of liquid and solid organic compounds. J. Photochem. Photobiol. A: Chem. **137**(1), 53–62 (2000)
9. Y. Ohko, K. Hashimoto, A. Fujishima, Kinetic analysis of the photocatalytic degradation of gas-phase 2-propanol under mass transport-limited conditions with a TiO_2 film photocatalyst. J. Phys. Chem. B **102**(10), 1724–1729 (1998)
10. J.H. Pan, H. Dou, Z. Xiong, C. Xu, J. Ma, X.S. Zhao, Porous photocatalysts for advanced waterpurifications. J. Mater. Chem. **20**(22), 4512–4528 (2010)
11. R. Wang, K. Hashimoto, A. Fujishima, M. Chikuni, E. Kojima, A. Kitamura, M. Shimohigoshi, T. Watanabe, Light-induced amphiphilic surfaces. Nature **388**, 431–432 (1997)
12. M. Miyauchi, A. Nakajima, A. Fujishima, K. Hashimoto, T. Watanabe, Photoinduced surface reactions on TiO_2 and $SrTiO_3$ films: photocatalytic oxidation and photoinduced hydrophilicity. Chem. Mater. **12**(1), 3–5 (2000)
13. R. Sun, A. Nakajima, A. Fujishima, T. Watanabe, K. Hashimoto, Photoinduced surface wettability conversion of ZnO and TiO_2 thin films. J. Phys. Chem. B **105**(10), 1984–1990 (2001)
14. N. Sakai, R. Wang, A. Fujishima, T. Watanabe, K. Hashimoto, Effect of ultrasonic treatment on highly hydrophilic TiO_2 surfaces. Langmuir **14**(20), 5918–5920 (1998)

15. M. Kamei, T. Mitsuhashi, Hydrophobic drawings on hydrophilic surfaces of single crystalline titanium dioxide: surface wettability control by mechanochemical treatment. Surf. Sci. **463**(1), L609–L612 (2000)
16. O. Ishiyama, T. Nishihara, S. Hayashi, M. Shinohara, M. Yoshimoto, T. Ohnishi, H. Koizumi, S. Nishino, J. Saraie, Atomic scale identification of the terminating structure of compound materials by CAICISS (coaxial impact collision ion scattering spectroscopy). Appl. Surf. Sci. **121–122**, 163–166 (1997)
17. M. Miyauchi, A. Nakajima, A. Fujishima, K. Hashimoto, T. Watanabe, A highly hydrophilic thin film under 1 $\mu W/cm^2$ UV illumination. Adv. Mater. **12**(24), 1923–1927 (2001)
18. M. Miyauchi, A. Nakajima, T. Watanabe, K. Hashimoto, Photoinduced hydrophilic conversion of TiO_2/WO_3 layered thin films. Chem. Mater. **14**(11), 4714–4720 (2002)
19. T. Shibata, A. Nakajima, A. Fujishima, T. Watanabe, K. Hashimoto, The 5th International Conference on TiO_2 Photocatalytic Purification and Treatment of Water and Air, London, Canada, Abstract of TiO_2–5, p.149 (2000)
20. A. Nakajima, S. Koizumi, T. Watanabe, K. Hashimoto, Effect of repeated photo-illumination on the wettability conversion of titanium dioxide. J. Photochem. Photobiol A. Chem. **146**(1–2), 129–132 (2001)
21. T. Nuida, N. Kanai, K. Hashimoto, T. Watanabe, H. Ohsaki, Enhancement of photocatalytic activity using UV light trapping effect. Vacuum **74**(3–4), 729–733 (2004)
22. R. Asahi, T. Morikawa, T. Ohwaki, K. Aoki, Y. Taga, Visible-light photocatalysis in nitrogen-doped titanium oxides. Science **293**(5528), 269–271 (2001)
23. T. Morikawa, R. Asahi, T. Ohwaki, K. Aoki, Y. Taga, Band-gap narrowing of titanium dioxide by nitrogen doping. Jpn. J. Appl. Phys. **40**(6A), L561–L563 (2001)
24. H. Irie, K. Sunada, K. Hashimoto, Recent developments in TiO_2 photocatalysis: novel applications to interior ecology materials and energy saving systems. Electrochemistry **72**(12), 807–812 (2004)
25. N. Sakai, A. Fujishima, T. Watanabe, K. Hashimoto, Enhancement of the photoinduced hydrophilic conversion rate of TiO_2 film electrode surfaces by anodic polarization. J. Phys. Chem. B **105**(15), 3023–3026 (2001)
26. J.M. White, J. Szanyi, M.A. Henderson, The photon-driven hydrophilicity of titania: A model study using TiO_2(110) and adsorbed trimethyl acetate. J. Phys. Chem. B **107**(34), 9029–9033 (2003)
27. C. Wang, H. Groenzin, M.J. Shultz, Molecular species on nanoparticulate anatase TiO_2 film detected by sum frequency generation: Trace hydrocarbons and hydroxyl groups. Langmuir **19**(18), 7330–7334 (2003)
28. R. Wang, K. Hashimoto, A. Fujishima, M. Chikuni, E. Kojima, A. Kitamura, M. Shimohigoshi, T. Watanabe, Photogeneration of highly amphiphilic TiO_2 surfaces. Adv. Mater. **10**(2), 135–138 (1998)
29. R. Wang, N. Sakai, A. Fujishima, T. Watanabe, K. Hashimoto, Studies of surface wettability conversion on TiO_2 single-crystal surfaces. J. Phys. Chem. B **103**(12), 2188–2194 (2000)
30. R. Nakamura, K. Ueda, S. Sato, In situ observation of the photoenhanced adsorption of water on TiO_2 films by surface-enhanced IR absorption spectroscopy. Langmuir **17**(8), 2298–2300 (2001)
31. A.Y. Nosaka, T. Fujiwara, H. Yagi, H. Akutsu, Y. Nosaka, Photoinduced changes of surface and adsorbed water in TiO_2 photocatalytic systems as studied by solid state ^{1}H-NMR spectroscopy. Chem. Lett. **31**(3), 420–421 (2002)
32. A.Y. Nosaka, E. Kojima, T. Fujiwara, H. Yagi, H. Akutsu, Y. Nosaka, Photoinduced changes of adsorbed water on a TiO_2 photocatalytic film as studied by ^{1}H NMR spectroscopy. J. Phys. Chem. B **107**(44), 12042–12044 (2003)
33. A.Y. Nosaka, T. Fujiwara, H. Yagi, H. Akutsu, Y. Nosaka, Photocatalytic reaction sites at the TiO_2 surface as studied by solid-state ^{1}H NMR spectroscopy. Langmuir **19**(6), 1935–1937 (2003)
34. A.Y. Nosaka, T. Fujiwara, H. Yagi, H. Akutsu, Y. Nosaka, Characteristics of water adsorbed on TiO_2 photocatalytic systems with increasing temperature as studied by solid-state ^{1}H NMR spectroscopy. J. Phys. Chem. B **108**(26), 9121–9125 (2004)
35. K. Uosaki, T. Yano, S. Nihonyanagi, Interfacial water structure at as-prepared and UV-induced hydrophilic TiO_2 surfaces studied by sum frequency generation spectroscopy and quartz crystal microbalance. J. Phys. Chem. B **108**(50), 19086–19088 (2004)
36. T. Shibata, H. Irie, K. Hashimoto, Enhancement of photoinduced highly hydrophilic conversion on TiO_2 thin films by introducing tensile stress. J. Phys. Chem. B **107**(39), 10696–10698 (2003)
37. A. Nakajima, S. Koizumi, T. Watanabe, K. Hashimoto, Photoinduced amphiphilic surface on polycrystalline anatase TiO_2 thin films. Langmuir **16**(17), 7048–7050 (2000)
38. K. Katsumata, A. Nakajima, H. Yoshikawa, T. Shiota, N. Yoshida, T. Watanabe, Y. Kameshima, K. Okada, Effect of microstructure on photoinduced hydrophilicity of transparent anatase thin films. Surf. Sci. **579**(2–3), 123–130 (2003)
39. M. Miyauchi, A. Nakajima, T. Watanabe K Hashimoto, Photocatalysis and photoinduced hydrophilicity of various metal oxide thin films, Chem. Mater. **14**(6), 2812–2816 (2002)
40. K. Katsumata, Y. Okayasu, N. Matsushita, K. Okada, Crystal-face dependence of photoinduced hydrophilic conversion on $SrTiO_3$ surfaces. Chem. Lett. **42**(6), 618–620 (2013)

41. K. Katsumata, C.E.J. Cordonier, T. Shichi, A. Fujishima, Photocatalytic activity of $NaNbO_3$ thin films. J. Am. Chem. Soc. **131**(11), 3856–3857 (2009)
42. K. Katsumata, S. Okazaki, C.E.J. Cordonier, T. Shichi, T. Sasaki, A. Fujishima, Preparation and characterization of self-cleaning glass for vehicle with niobia nanosheets. ACS Appl. Mater. Interfaces **2**(4), 1236–1241 (2010)
43. F. Gyntelberg, P. Suadicani, J.W. Nielsen, P. Skov, O. Valbjørn, P.A. Nielsen, T. Schneider, O. Jørgensen, P. Wolkoff, C. Wilkins, Dust and the sick building syndrome. Indoor Air **4**(4), 223–238 (1994)
44. R. Bholah, A. Subratty, Indoor biological contaminants and symptoms of sick building syndrome in office buildings in Mauritius. Int. J. Environ. Health Res. **12**(1), 93–98 (2002)
45. J.M. Boyce, Environmental contamination makes an important contribution to hospital infection. J. Hosp. Infect. **65**, 50–54 (2007)
46. J.S. Peiris, K.Y. Yuen, A.D. Osterhaus, K. Stöhr, The severe acute respiratory syndrome. N. Engl. J. Med. **349**(25), 2431–2441 (2003)
47. R. Maunder, J. Hunter, L. Vincent, J. Bennett, N. Peladeau, M. Leszcz, J. Sadavoy, L.M. Verhaeghe, R. Steinberg, T. Mazzulli, The immediate psychological and occupational impact of the 2003 SARS outbreak in a teaching hospital. Can. Med. Assoc. J. **168**(10), 1245–1251 (2003)
48. C. Mashonganyika, H. McKee, J. Board, A. Davies, S. Vallance, F. Hatch, P. Stewart, A. McKee, S. McGuinness, R. Parke, Critical care services and 2009 H1N1 influenza in Australia and New Zealand. N. Engl. J. Med. **361**(20), 1925–1934 (2009)
49. A. Macfarlane, Daily mortality and environment in English conurbations. Air pollution, low temperature, and influenza in Greater London. J. Epidemiol. Community Health, **31** (1), 54–61 (1977)
50. P.N. Dinh, H.T. Long, N.T.K. Tien, N.T. Hien, L.T.Q. Mai, L.H. Phong, L. Van Tuan, H. Van Tan, N.B. Nguyen, P. Van Tu, Risk factors for human infection with avian influenza A H5N1, Vietnam, 2004. Emerg. Infect. Dis. **12**(12), 1841 (2006)
51. J. T. Park, J. L. Strominger, Mode of action of penicillin. Biochemical basis for the mechanism of action of penicillin and for its selective toxicity. Science. **125** (3238), 99–101 (1957)
52. J.L. Strominger, J.T. Park, R.E. Thompson, Composition of the cell wall of Staphylococcus aureus: its relation to the mechanism of action of penicillin. J. Biol. Chem. **234**(12), 3263–3268 (1959)
53. B. Setlow, C. Loshon, P. Genest, A. Cowan, C. Setlow, P. Setlow, Mechanisms of killing spores of Bacillus subtilis by acid, alkali and ethanol. J. Appl. Microbiol. **92**(2), 362–375 (2002)
54. A.J. Hall, J. Vinjé, B. Lopman, G.W. Park, C. Yen, N. Gregoricus, U. Parashar, Updated norovirus outbreak management and disease prevention guidelines. Morb. Mortal. Wkly Rep. **60**(3), 1–15 (2011)
55. M. Khadre, A. Yousef, J.G. Kim, Microbiological aspects of ozone applications in food: a review. J. Food Sci. **66**(9), 1242–1252 (2001)
56. K. Nakata, A. Fujishima, TiO2 photocatalysis: design and applications. J. Photochem. Photobiol. C **13**(3), 189 (2012)
57. K. Nakata, T. Ochiai, T. Murakami, A. Fujishima, Photoenergy conversion with TiO2 photocatalysis: new materials and recent applications. Electrochim. Acta. **84**, 111 (2012)
58. A. Fujishima, K. Nakata, T. Ochiai, A. Manivannan, D.A. Tryk, Recent aspects of photocatalytic technologies for solar fuels, self-cleaning, and environmental cleanup. Interface **22**(2), 56 (2013)
59. A. Fujishima, X. Zhang, D.A. Tryk, TiO_2 photocatalysis and related surface phenomena. Surf. Sci. Rep. **63**(12), 515–582 (2008)
60. I. Sopyan, M. Watanabe, S. Murasawa, K. Hashimoto, A. Fujishima, An efficient TiO_2 thin-film photocatalyst: photocatalytic properties in gas-phase acetaldehyde degradation. J. Photochem. Photobiol., A **98**(1), 79–86 (1996)
61. H. Einaga, S. Futamura, T. Ibusuki, Heterogeneous photocatalytic oxidation of benzene, toluene, cyclohexene and cyclohexane in humidified air: comparison of decomposition behavior on photoirradiated TiO_2 catalyst. Appl. Catal. B **38**(3), 215–225 (2002)
62. T. Matsunaga, R. Tomoda, T. Nakajima, H. Wake, Photoelectrochemical sterilization of microbial cells by semiconductor powders. FEMS Microbiol. Lett. **29**(1–2), 211–214 (1985)
63. J.C. Ireland, P. Klostermann, E.W. Rice, R.M. Clark, Inactivation of Escherichia coli by titanium dioxide photocatalytic oxidation. Appl. Environ. Microbiol. **59**(5), 1668–1670 (1993)
64. M. Cho, H. Chung, W. Choi, J. Yoon, Different inactivation behaviors of MS-2 phage and escherichia coli in TiO2 photocatalytic disinfection. Appl. Environ. Microbiol. **71**(1), 270–275 (2005)
65. S. Lee, K. Nishida, M. Otaki, S. Ohgaki, Photocatalytic inactivation of phage Qβ by immobilized titanium dioxide mediated photocatalyst. Water Sci. Technol. **35**(11), 101–106 (1997)
66. J. Lonnen, S. Kilvington, S. Kehoe, F. Al-Touati, K. McGuigan, Solar and photocatalytic disinfection of protozoan, fungal and bacterial microbes in drinking water. Water Res. **39**(5), 877–883 (2005)
67. R. Nakano, M. Hara, H. Ishiguro, Y. Yao, T. Ochiai, K. Nakata, T. Murakami, J. Kajioka, K. Sunada, K. Hashimoto, Broad spectrum microbicidal activity of photocatalysis by TiO2. Catalysts **3**(1), 310–323 (2013)
68. R.J. Watts, S. Kong, M.P. Orr, G.C. Miller, B.E. Henry, Photocatalytic inactivation of coliform bacteria and viruses in secondary wastewater effluent. Water Res. **29**(1), 95–100 (1995)

69. O.K. Dalrymple, E. Stefanakos, M.A. Trotz, D.Y. Goswami, A review of the mechanisms and modeling of photocatalytic disinfection. Appl. Catal. B **98**(1), 27–38 (2010)
70. H.A. Foster, I.B. Ditta, S. Varghese, A. Steele, Photocatalytic disinfection using titanium dioxide: spectrum and mechanism of antimicrobial activity. Appl. Microbiol. Biotechnol. **90**(6), 1847–1868 (2011)
71. G. Gogniat, S. Dukan, TiO2 photocatalysis causes DNA damage via Fenton reaction-generated hydroxyl radicals during the recovery period. Appl. Environ. Microbiol. **73**(23), 7740–7743 (2007)
72. G. Carré, E. Hamon, S. Ennahar, M. Estner, M.-C. Lett, P. Horvatovich, J.-P. Gies, V. Keller, N. Keller, P. Andre, TiO2 photocatalysis damages lipids and proteins in Escherichia coli. Appl. Environ. Microbiol. **80**(8), 2573–2581 (2014)
73. Y. Kakita, N. Kashige, F. Miake, K. Watanabe, Photocatalysis-dependent inactivation of Lactobacillus phage PL-1 by a ceramics preparation. Biosci. Biotechnol. Biochem. **61**(11), 1947–1948 (1997)
74. L. Belháčová, J. Krýsa, J. Geryk, J. Jirkovský, Inactivation of microorganisms in a flow-through photoreactor with an immobilized TiO2 layer. J. Chem. Technol. & Biotechnol.: Int. Res. Process., Environ. & Clean Technol. **74**(2), 149–154 (1999)
75. J.C. Sjogren, R.A. Sierka, Inactivation of phage MS2 by iron-aided titanium dioxide photocatalysis. Appl. Environ. Microbiol. **60**(1), 344–347 (1994)
76. D. Gerrity, H. Ryu, J. Crittenden, M. Abbaszadegan, Photocatalytic inactivation of viruses using titanium dioxide nanoparticles and low-pressure UV light. J. Environ. Sci. Health Part A **43**(11), 1261–1270 (2008)
77. M.V. Liga, E.L. Bryant, V.L. Colvin, Q. Li, Virus inactivation by silver doped titanium dioxide nanoparticles for drinking water treatment. Water Res. **45**(2), 535–544 (2011)
78. N. Kashige, Y. Kakita, Y. Nakashima, F. Miake, K. Watanabe, Mechanism of the photocatalytic inactivation of Lactobacillus casei phage PL-1 by titania thin film. Curr. Microbiol. **42**(3), 184–189 (2001)
79. H. Ishiguro, R. Nakano, Y. Yao, J. Kajioka, A. Fujishima, K. Sunada, M. Minoshima, K. Hashimoto, Y. Kubota, Photocatalytic inactivation of bacteriophages by TiO_2-coated glass plates under low-intensity, long-wavelength UV irradiation. Photochem. Photobiol. Sci. **10**(11), 1825–1829 (2011)
80. J.-W. Liou, H.-H. Chang, Bactericidal effects and mechanisms of visible light-responsive titanium dioxide photocatalysts on pathogenic bacteria. Arch. Immunol. Ther. Exp. (Warsz.) **60**(4), 267–275 (2012)
81. M. Liu, K. Sunada, K. Hashimoto, M. Miyauchi, Visible-light sensitive Cu (II)–TiO_2 with sustained anti-viral activity for efficient indoor environmental remediation. J. Mater. Chem. A **3**(33), 17312–17319 (2015)
82. K. Takehara, K. Yamazaki, M. Miyazaki, Y. Yamada, S. Ruenphet, A. Jahangir, D. Shoham, M. Okamura, M. Nakamura, Inactivation of avian influenza virus H1N1 by photocatalyst under visible light irradiation. Virus Res. **151**(1), 102–103 (2010)
83. R. Nakano, H. Ishiguro, Y. Yao, J. Kajioka, A. Fujishima, K. Sunada, M. Minoshima, K. Hashimoto, Y. Kubota, Photocatalytic inactivation of influenza virus by titanium dioxide thin film. Photochem. Photobiol. Sci. **11**(8), 1293–1298 (2012)

Environmental Degradation of Polymers and Methods of Its Acceleration/Suppression

Nhan Nu Thanh Ton, Anh Thi Ngoc Dao, Kalaivani Seenivasan, Emi Sawade and Toshiaki Taniike

1 Environmental Degradation of Polymers

The degradation of polymers is a process in which polymers exposed to external factors for a long period of time undergo physical and/or chemical changes [1], which causes the alternation of their original properties like changing in color (e.g., yellowing), texture, mechanical properties, and so on [2, 3]. The degradation of polymers occurs during processing, service, exposure to the environment, recycling, etc., by various mechanisms, which include mechanical, thermal, oxidative, chemical, hydrolytic, photochemical, and biological ones. Among all, the degradation of polymers in the environment is currently the primary concern of scientists and the subject of this chapter, as plastic wastes are causing serious impacts on ecosystems, especially marine life.

Depending on the chemical structure of polymers and the environment to which they are exposed, the mechanism of the environmental degradation and important factors therein can vary. Abiotic degradation occurs in mechanical, thermal, oxidative, chemical, hydrolytic, and photochemical manners. Mechanical degradation can be caused by disintegration and abrasion forces resulting from the collision of plastic wastes with pebbles, sediments, vessels, and animals in the aquatic environment, or from the activities of human and animals in the terrestrial environment. In addition, natural phenomena like tornadoes, rainstorms, tidal flow, in particular, temperature or humidity differences largely contribute to the mechanical degradation of polymers [4]. The state of polymers, *i.e.,* solid (crystalline or glassy) or rubbery, is another key factor in the mechanical degradation pathway. For polymers in a solid state, the formation of voids and fracture planes might induce the breakage of chemical bonds. In a rubbery state, chain entanglement leads to the stretching of a part of the macromolecules, which can cause the breakage of chemical bonds. In environmental degradation, the mechanical degradation plays an essential role in initiating and developing cracks

N. N. T. Ton · K. Seenivasan · E. Sawade · T. Taniike (✉)
Graduate School of Advanced Science and Technology, Japan Advanced Institute of Science and Technology, 1-1 Asahidai, Nomi 923-1292, Ishikawa, Japan
e-mail: taniike@jaist.ac.jp

N. N. T. Ton
e-mail: ton-ntn@jaist.ac.jp

K. Seenivasan
e-mail: kalaivan@jaist.ac.jp

E. Sawade
e-mail: e-sawade@jaist.ac.jp

A. T. N. Dao
Graduate School of Engineering, Nagasaki University, 1-14 Bunkyo-machi, Nagasaki 852-8521, Japan
e-mail: anh.dao@nagasaki-u.ac.jp

T. Kaneko (ed.), *Photo-switched Biodegradation of Bioplastics in Marine Environments*,
https://doi.org/10.1007/978-981-99-4354-8_6

in polymers and accelerating the overall degradation process. The thermal mechanism is not a direct cause of the environmental degradation, but it should be noted that the temperature has a significant impact on the degradation through activation processes and polymer mobility [5, 6]. Oxidative degradation, on the other hand, is an important mechanism in the environmental degradation since oxygen is the most abundant atmospheric component. Oxidative degradation can be accelerated by the action of microorganisms, light, heat, or combination of them [7, 8]. For example, under solar light irradiation, oxidation can occur concomitantly or synergistically to photodegradation to generate free radicals which subsequently lead to crosslinking reactions or chain scission. Chemical stress is also an essential abiotic factor in the environmental degradation of polymers, in which the interaction of agrochemicals with polymers can change macromolecular properties [9]. Besides, the interaction of polymers with air pollutants, in particular NO_2 and SO_2, is also concerned [10–13]. It was reported that unsaturated polymers and polyamides are reactive with NO_2 and SO_2, causing chain scission and crosslinking [10–13], which are accelerated under photoirradiation. In addition, the environmental degradation of polymers can occur through hydrolysis in the presence of water [14, 15], which is important when the polymer contains hydrolyzable covalent bonds like ester, anhydride, amine, ether, or carbamide groups. The rate of hydrolysis depends on the molecular weight, the sensitivity of the chemical bonds of the polymers to water, its concentration inside the polymer, and water diffusion in the material [16].

The last and the most essential abiotic factor in the environmental degradation of a polymer is solar light. This degradation process is called photodegradation. A majority of polymers undergo degradation reactions under irradiation, particularly in sunlight due to the existence of chromophoric groups in the macromolecules or residual impurities. Photodegradation can undergo mainly through Norrish reactions and/or crosslinking in the absence of oxygen, while through photooxidative degradation in the presence of oxygen [8, 17]. Among these, the photooxidative degradation is the most common mode of the degradation in the environment. The photooxidative degradation consists of three main steps: photoinitiation, propagation, and termination [18–20]. In the photoinitiation, chromophoric groups (initiators) absorb light and cleave to radicals. Hydroperoxides and ketones are the most important initiators [21–23]. Hydroperoxides are photolabile; they can be easily decomposed to form radicals, which can then abstract hydrogen atoms in the polymer to initiate the photooxidation process. Alkoxy radicals produced from the photolysis of hydroperoxides cause β-scission to form carbonyl groups [22]. Ketones absorb photons, and directly lead to the scission of the polymer backbone through C–C bond breakage [24, 25]. For example, the photodegradation of polyethylene and polypropylene is initiated by hydroperoxides and/or ketonic oxidation products introduced during melt processing [26–28]. In the case of polyamide, the abstraction of the hydrogen atom in methylene groups adjacent to the nitrogen atom of CO–NH bonds is the initiation step of the photodegradation process [29–31]. The next step, the propagation, consists of (i) reactions between low molecular weight radicals and macromolecules; (ii) reactions of macromolecular radicals with oxygen molecules to form peroxy radicals (ROO·); (iii) the abstraction of a hydrogen atom from macromolecules by ROO· to form a hydroperoxide (ROOH); (iv) the photolysis of ROOH into alkoxy (RO·) and hydroxyl (HO·) radicals; (v) the abstraction of a hydrogen by RO· to form ROH; (vi) β-scission of RO· into aldehyde end groups and macromolecular alkyl radicals [32]. The final step is termination, in which radicals formed in the polymer photodegradation are terminated by different combination reactions with the formation of inactive products.

The other important type of the environmental degradation of polymers is biological degradation, which is caused by the activity of microorganisms (e.g., bacteria, algae, fungi, protozoa, and lichenaceae groups) on the surface and/or inside a polymeric material [33–35].

The development of microorganisms depends on the nature of polymers and also on environmental conditions, e.g., temperature, humidity, and atmospheric pollutants as potential nutrient sources for some microbial species [36–40]. Mixed populations of microorganisms can adhere to polymer surfaces in organized consortia called biofilms [41], in which they work synergistically with each other, causing serious damage to the polymers [42–45]. Biodeterioration of polymers can take place in physical, chemical, or enzymatic manners [41]. Physically, microorganisms can adhere to polymer surfaces via secretion of a kind of glue and form a slime biofilm [46], which penetrates a porous structure of the polymers, increases the pore size, and induces cracks [47]. Chemically, extracellular polymers formed by microorganisms can present as surfactants, which assist the penetration of microorganisms. Besides, the existence of slime facilitates the accumulation of atmospheric pollutants, thus increasing the development of microbial species and accelerating the biodeterioration [37]. The last biodeterioration pathway is enzymatic one, which is the most important in the biodeterioration of polymers, especially of high molecular weight [48]. Enzymes that are involved in the biodeterioration of polymers are mostly clarified into two categories: extracellular and intracellular depolymerases [41]. The enzymatic biodeterioration is initiated by the activation of extracellular depolymerases, which are also called exoenzymes. With the wide range of activity against reactions including oxidation and hydrolysis, exoenzymes can degrade a polymer structure to smaller units, which can be then taken into microbial cells for complete degradation with the aid of intracellular enzymes [49].

It can be seen that the environmental degradation of polymers is a long-term and complex process resulting from the interaction of many factors such as those introduced above. In fact, controlling the environmental degradation of polymers is still a major challenge today. Accelerating the degradation into unharmful forms is required when a polymer waste is exposed to the environment, while practical use requires polymers to equip the least stability under environmental factors. A key to reconciling the conflicting properties, the environmental degradability and stability of polymers, is the combination of catalysts that promote degradation and stabilizers that inhibit degradation, as well as controlling their functions according to environmental factors during use and disposal.

2 Photocatalysts in Accelerating the Environmental Degradation of Polymers

The environmental degradation of polymers can be accelerated by several strategies. For example, blending synthetic polymers with natural polymers like starch [50, 51], chitosan [52, 53], or with natural fibers [54, 55] have been widely studied in order to enhance the biodegradability of the products, in which major obstacles are deterioration of mechanical properties [56, 57], high moisture sensitivity [56, 58], and interfacial failure [59]. To enhance the hydrophilicity of polymers by additives like polyethylene glycol [60, 61], graphene oxide [62, 63], or hydrophobins [64] and it also has the potential to promote the adherence of microbials to the polymer surfaces, which in turn accelerates the biodegradation. The addition of catalysts is an effective strategy to accelerate environmental degradation: enzymes for accelerated biodegradation [65, 66], and more commonly, photocatalysts for accelerated photodegradation [67–73]. The photodegradation plays a significant role in the environmental degradation of polymers since sunlight is an essential abiotic factor unless polymer wastes are landfilled. Hence, the development of photocatalysts, especially those with visible light response and compatibility with polymers, is a major concern for the acceleration of the environmental degradation of polymers. What follows describes our photocatalyst technology and its use to promote the photodegradation of polymers.

A variety of photocatalysts have been employed for the purpose of accelerated polymer photodegradation, which includes TiO_2 [67, 68], ZnO [69, 70], SnO_2 [71], CN_x/Ni_2P [72], CdS/CdO_x[73], and so on. Among these, TiO_2 is the most commonly employed photocatalyst owing to its low cost, high oxidation capability, high physicochemical and photostability, and non-corrosive behavior [74–78]. For example, TiO_2 can be added into biodegradable films for food packaging not only as an ethylene scavenger to prolong the shelf life of fresh products and but also as a photocatalyst to accelerate the degradation of the polymer after service [79, 80]. In the agricultural field, photodegradable polymer nanocomposites with TiO_2 have widely been investigated, which is represented by low-density polyethylene (LDPE)/TiO_2 [81–83]. Besides these successful applications, the large bandgap and short lifetime of charge carriers of TiO_2 are known as the main drawbacks in visible light photocatalysis [74, 84]. Besides, the hydrophilic nature of TiO_2 humpers uniform dispersion of the particles in polymer matrices. Many efforts have been devoted to enhancing the visible light photocatalytic performance of TiO_2, for example, by doping with metal ion or non-metals [85–90], grafting with chelating ligands [91–93], or tailoring heterojunction with other semiconductors [94, 95]. Pairing with graphene is considered as one of the most effective approaches to overcome the shortfalls of TiO_2. With an appropriate work function and the 2D π-conjugate structure, graphene can work as an electron sink, which impedes the recombination of photoinduced charge carriers [96, 97]. The hydrophobic nature of graphene is also advantageous in dispersing catalyst particles in polymer matrices.

A few challenges have been pointed out in the synthesis of TiO_2/graphene composites, which are rooted in the use of graphene oxide (GO) as a starting material. The reduction of GO can generate a large amount of defects in the graphene framework, which adversely affects the electronic properties of formed graphene, including hindered ballistic transport [98, 99]. Moreover, water molecules, which GO usually contains, react with water-sensitive titanium alkoxide and obstruct the uniform growth of TiO_2 on graphene sheets [100–104]. Recently, we established a novel GO-free route to fabricate TiO_2/graphene nanocomposites using a graphene dispersion in titanium alkoxide as a starting material [105]. We found that chemical exfoliation of graphite in titanium tetra-*n*-butoxide ($Ti(OnBu)_4$) under ultrasonication yields a dispersion of high-quality graphene, which was confirmed by Raman spectroscopy. Figure 1a compares the Raman spectra of graphene produced by chemical exfoliation of graphite using $Ti(OnBu)_4$ as an exfoliating solvent, and that produced using benzylamine as a common exfoliating solvent [106–108]. In general, the Raman spectrum of graphene is featured with the G band, which arises from the in-plane vibration of sp^2 carbon atoms [109], and the D band, which is attributed to the existence of defects in the graphene framework [110]. The amount of defects with respect to the sp^2 hybridized graphene domains is evaluated by the intensity ratio of the D and G band (I_D/I_G) [111]. In addition, the full width at half maximum (FWHM) of the 2D two-phonon band is known to correlate well with the number of graphene layers [112]. As shown in Fig. 1a, the I_D/I_G ratio of graphene produced in the presence of $Ti(OnBu)_4$ was 0.20, which was much smaller than 0.69 when benzylamine was used, 0.67 for graphene prepared by electrochemical exfoliation of graphite [113, 114], and 0.87 for reduced GO [115–117]. Besides, a sharp 2D band with a low value of FWHM (28 cm^{-1}) dictated the formation of few-layer graphene in $Ti(OnBu)_4$ [118], while the use of benzylamine led to multi-layer graphene with an FWHM value of 70 cm^{-1} [118]. The number of graphene layers was also investigated by atomic force microscopy (AFM), in which the thickness of the exfoliated graphene layers on a SiO_2/Si substrate was recorded. It is known that the thickness of single-layer graphene (SLG) on a SiO_2/Si substrate is around 1 nm [119]. Thus, the number of graphene layers can be determined by dividing the measured

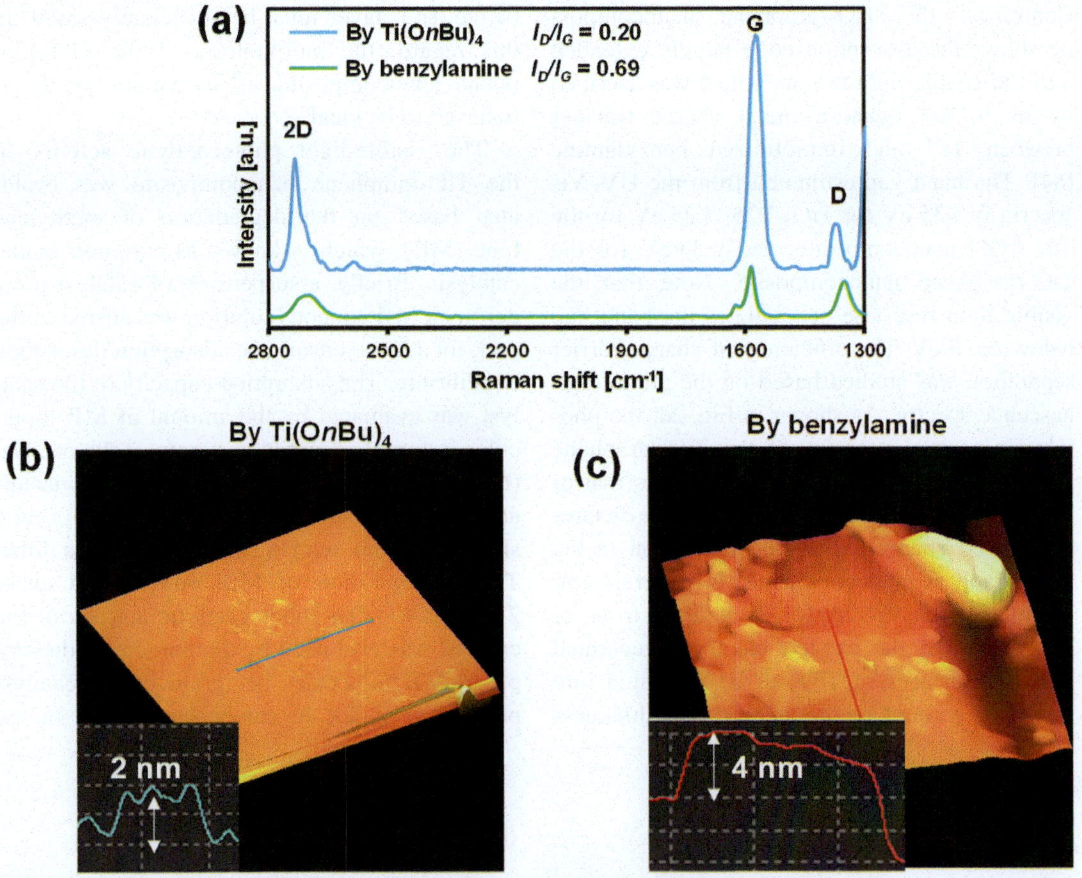

Fig. 1 **a** Raman spectra of graphene prepared by chemical exfoliation of graphite using $Ti(OnBu)_4$ as an exfoliating solvent, and that prepared using benzylamine. **b**, **c** AFM images of the two graphene samples

thickness with the SLG thickness. Figure 1b, c shows AFM images of graphene prepared using $Ti(OnBu)_4$ and benzylamine as exfoliating solvents, respectively. The thickness of graphene sheets of around 2 nm confirmed the formation of bi-layer graphene in $Ti(OnBu)_4$, whereas thicker graphene (four layers) was produced in benzylamine. Likewise, the graphene dispersion in $Ti(OnBu)_4$ is considered as an ideal precursor for the fabrication of TiO_2/graphene nanocomposites. This is implemented via a sol–gel reaction, for example, by dropping an aqueous solution of ammonia and benzylamine into the dispersion.

A TiO_2/graphene nanocomposite prepared from the graphene dispersion in $Ti(OnBu)_4$ equips several important characteristics, which are not presented in conventional nanocomposites prepared using GO as a starting material. For example, the TiO_2/graphene nanocomposite possesses a sheet-like morphology, in which isolated graphene sheets are uniformly covered by an nm-thick TiO_2 layer (Fig. 2a). Meanwhile, the TiO_2/rGO reference nanocomposite, which was synthesized based on the conventional solvothermal method using GO and $Ti(OnBu)_4$ as starting materials, consists of agglomerated granular TiO_2 particles of *ca.* 10–20 nm (Fig. 2b). Pristine TiO_2 and the TiO_2/rGO nanocomposite hardly exhibit optical absorption in the visible light region (Fig. 2c), i.e., their visible light response is expected to be poor.

Contrarily, the TiO_2/graphene nanocomposite shows the absorption edge largely extended into the visible light region, which was ascribed by us to the ligand-to-metal charge transfer between Ti^{4+} and immobilized benzylamine [84]. The band gap estimated from the UV–Vis spectra is 3.35 eV for TiO_2-P25, 3.25 eV for the TiO_2/rGO nanocomposite, and 2.33 eV for the TiO_2/graphene nanocomposite. Note that the visible light response necessitates the band gap below *ca.* 3 eV. The efficiency of charge carrier separation was studied based on the photoluminescence spectra. As shown in Fig. 2d, the photoluminescence intensity of the TiO_2/graphene nanocomposite is far lower compared to that of TiO_2-P25 and TiO_2/rGO, which directly dictates suppressed charge carrier recombination in the TiO_2/graphene nanocomposite. In general, graphene works as a sink for excited electrons of TiO_2, and for the excited electrons generated near the surfaces of TiO_2 to be separated into graphene without recombination, the thickness of the TiO_2 layer must be sufficiently small. In this regard, the morphology of the TiO_2/graphene nanocomposite, as shown in Fig. 2a, is believed to be ideal.

The visible-light photocatalytic activity of the TiO_2/graphene nanocomposite was evaluated based on the degradation of methylene blue (MB), which is known as common model catalysis. Briefly, a suspension of catalyst powder in an MB aqueous solution was stirred in the dark for 12 h to ensure the adsorption/desorption equilibrium. The adsorption capacity of the catalyst was evaluated by the amount of MB molecules in mmol adsorbed per gram of the catalyst (Fig. 3a). Thereafter, the suspension was irradiated at room temperature using a 300 W xenon short arc lamp with a 422-nm longpass filter. The concentration of MB was tracked every 2 h up to 6 h. The photocatalytic activity of the catalyst was evaluated by the amount of decomposed MB molecules per gram of the catalyst per hour (Fig. 3b). As can be seen in Fig. 3a, the

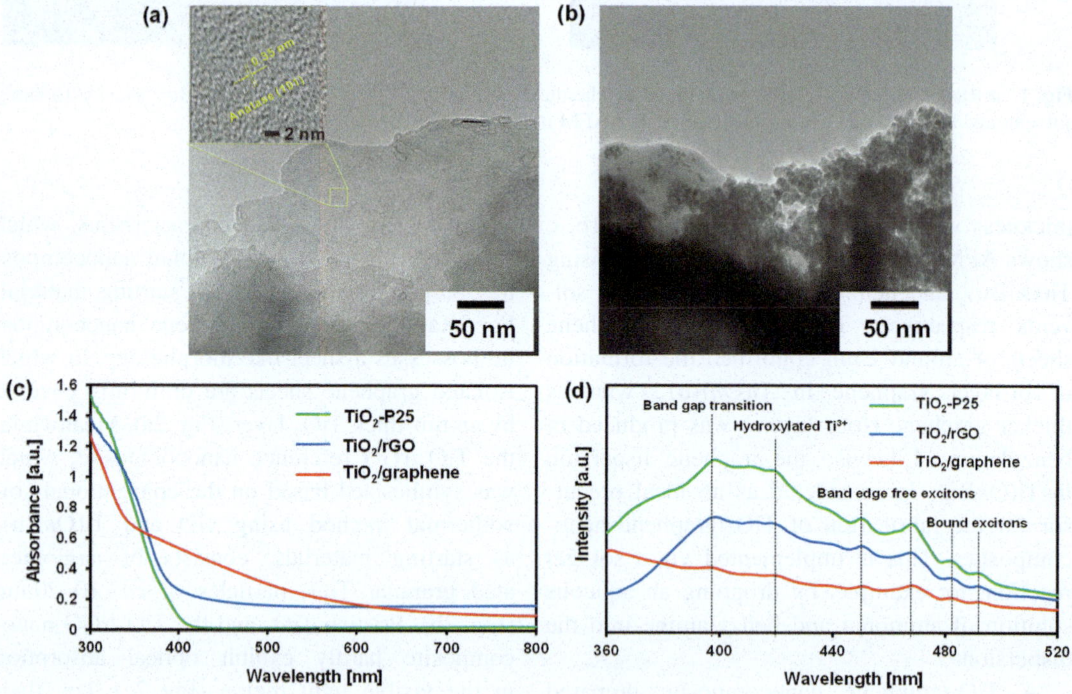

Fig. 2 TEM images of **a** TiO_2/graphene and **b** TiO_2/rGO nanocomposites. **c** UV–Vis diffuse reflectance and **d** photoluminescence spectra of the two nanocomposites together with the spectra of TiO_2 (P25) as a reference

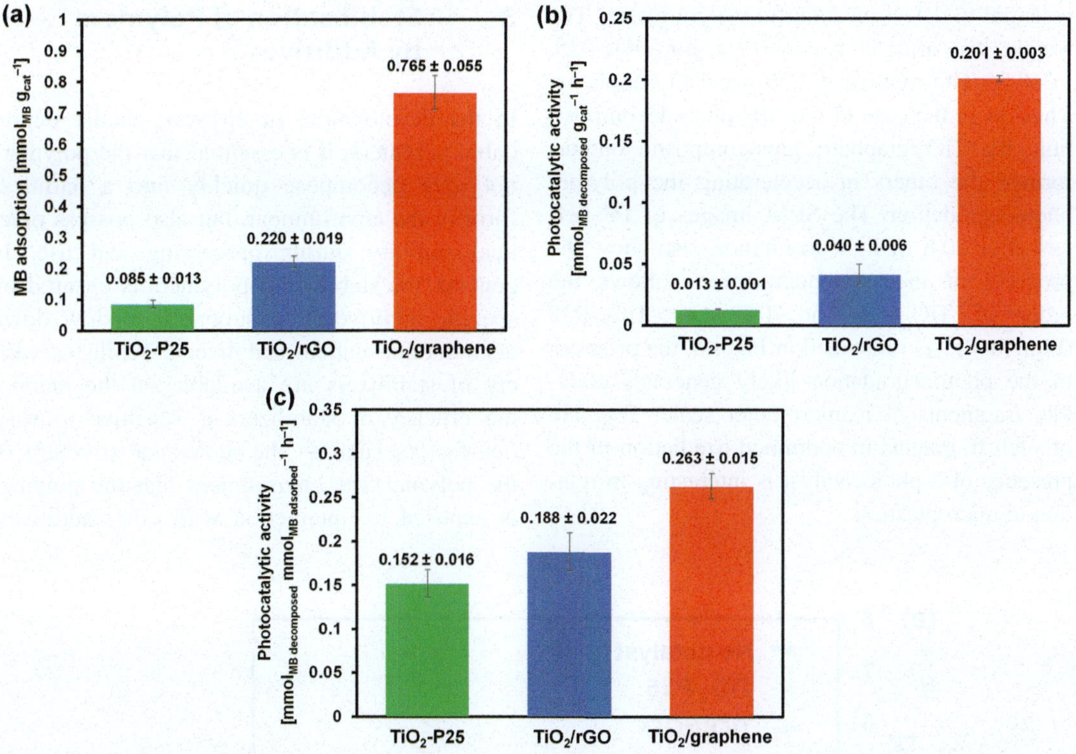

Fig. 3 **a** MB adsorption capacity in dark, **b** photocatalytic activity per gram catalyst, **c** photocatalytic activity per the amount of MB molecules adsorbed

incorporation of graphene increased the adsorption capacity of the catalysts compared to that of TiO_2-P25 due to enlarged surface area and/or enhanced affinity towards conjugated molecules. Noticeably, the TiO_2/graphene nanocomposite exhibited far the highest adsorption capacity based on the highest surface area of the material. In Fig. 3b, the visible light photocatalytic activity of the TiO_2/graphene nanocomposite was found far the greatest, *i.e.*, 15 and 5 times higher than the activity of TiO_2-P25 and the TiO_2/rGO nanocomposite, respectively. In order to understand the origin of the excellent photocatalytic activity of the TiO_2/graphene nanocomposite, the activity per gram of the catalyst was converted into that per mmol of MB molecules adsorbed (Fig. 3c). After such normalization, the activity of the TiO_2/graphene nanocomposite over TiO_2/rGO became 1.4 times, in contrast to the 5 times before the normalization. Thus, a large portion (*ca.* 70%) of the improvement in the activity of the TiO_2/graphene nanocomposite is attributed to its excellent adsorption capacity, as shown in Fig. 3a. The remaining portion of *ca.* 30% is explained by either of the extension of the absorption edge into the visible light region and effective charge carrier separation benefited from the thin TiO_2 layer covering the defect-less graphene sheets.

Following the success in the model catalysis, the developed catalyst was tested in the photodegradation of a polymer, where polypropylene (PP) was selected as one of the most widely used plastics. In brief, a fixed amount of catalyst powder was suspended in DI water under vigorous stirring for 1 h, to which PP pellets were added. The mixture was subjected to visible light irradiation at room temperature, where the degradation was tracked based on the weight loss of the PP pellets every 4 h up to 20 h (Fig. 4a). The self-photodecomposition of PP was negligible within 20 h of irradiation; the presence of catalysts was essential to detect meaningful weight losses. The weight loss after 20 h was 0.8% for TiO_2-P25, 1.3% for TiO_2/rGO, and 7.5% for TiO_2/graphene. Thus, as in the case of the MB photodecomposition, the TiO_2/graphene nanocomposite outperformed the others in accelerating the polymer photodegradation. The SEM images of PP pellets after 20 h of light irradiation also show the progress of the photodegradation follows the order of TiO_2/graphene > TiO_2/rGO > TiO_2-P25 (Fig. 4b–d). As illustrated in Fig. 4d, the progress of the photodegradation likely generates plate-like fragments of a micrometer scale. The fate of such fragments in continued irradiation in the presence of a photocatalyst is interesting in relation to microplastics.

3 Stabilization of Polymers by Additives

In the development of environmentally degradable polymers, it is essential that the polymers not only decompose quickly into a harmless form in the environment, but also possess practical stability during processing and use. In general, the stability of polymers is ensured by specific additives that suppress or slow down degradation, called stabilizers. While a variety of stabilizers are available on the market, the efficacy of stabilizers is sensitive to various factors such as the molecular structure of the polymer, the environment that the polymer is exposed to, interaction with other additives,

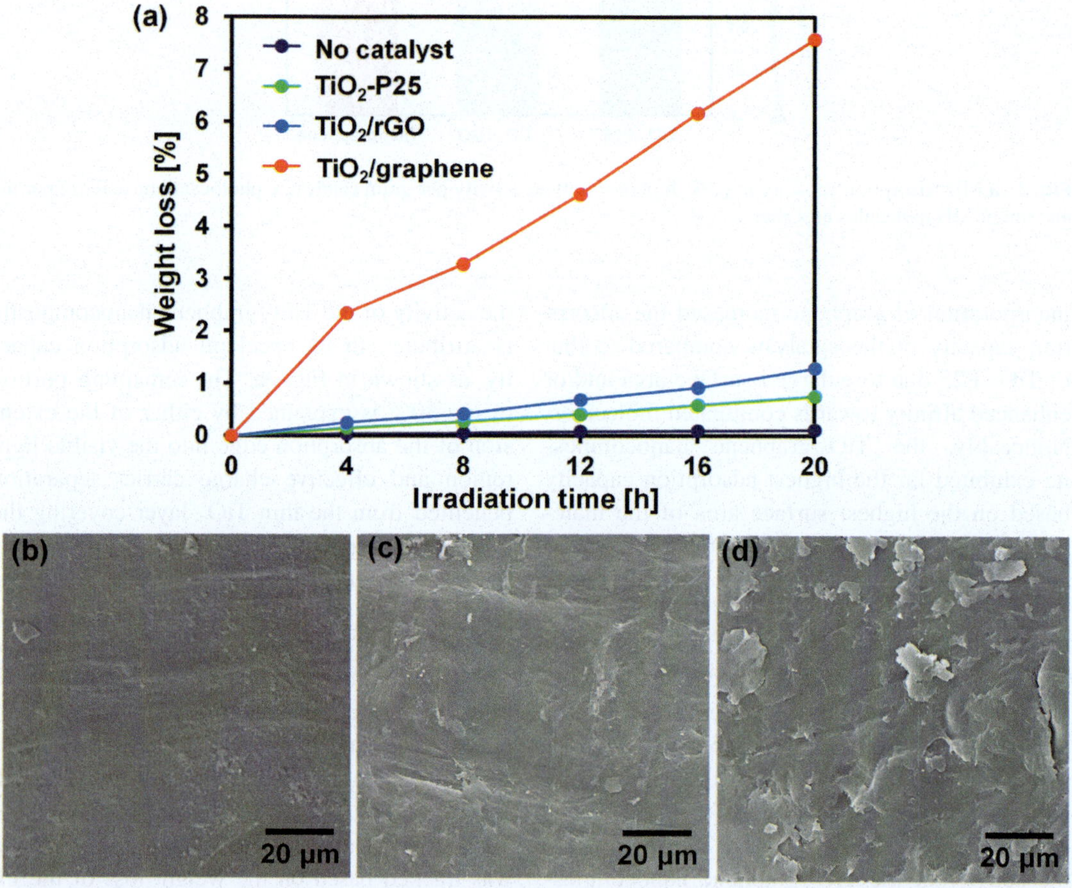

Fig. 4 Photodegradation of PP pellets in water in the presence of photocatalysts. **a** The progress of the photodegradation was monitored by the weight loss. SEM images of PP pellets irradiated for 20 h in the presence of **b** TiO_2-P25, **c** TiO_2/rGO, and **d** TiO_2/graphene

and so on. Therefore, it is inevitable to pick up appropriate stabilizers in each case, and this can only be done in an empirical manner. We have developed high-throughput experimental methods and used them in screening stabilizers for conventional plastics such as PP [120, 121] as well as for emerging bio-based plastics [122, 123]. What follows describes our efforts on the stabilization of recombinant spider silk against the thermo-oxidative degradation [122].

The thermo-oxidative degradation of polymers often accompanies the emission of chemiluminescence (CL), and measuring the CL emission has been recognized as one of the most sensitive techniques to detect the oxidative degradation of polymers [120–122, 124]. In particular, the change of the CL intensity along the degradation time, called a CL curve, is subject to evaluate the efficacy of antioxidants. We have developed a high-throughput chemiluminescence imaging (HTP-CLI) instrument, which enables us to simultaneously acquire the CL curves for 100 polymer samples within a single measurement (Fig. 5a). This instrument was used to screen various antioxidants in the thermo-oxidative degradation of recombinant spider silk. The tested antioxidants include nature-derived ones, hindered phenols, thioethers, and phosphites (Scheme 1).

The CL behavior is highly dependent on the structure of polymers and the mechanism of oxidative degradation. For example, the CL emission in the oxidative degradation of recombinant spider silk arises from the unimolecular decomposition of unstable hydroperoxides generated at C_α of amino acid residues [125], where the emission gradually increases from the beginning of aging and then decreases after reaching a maximum. As illustrated in Fig. 5b, the deceleration in the oxidation delays the time for the emission to reach the maximum intensity (t_{Imax}), i.e., the efficacy of antioxidants can be evaluated based on t_{Imax}. In the example shown in Fig. 6a, HTP-CLI measurements were performed at 200 °C for recombinant spider silk powder impregnated with individual antioxidants of 0.3 wt%. The t_{Imax} of each sample was determined from the CL curve and it was converted into Δt_{Imax}, which corresponds to the percentage increase or decrease from t_{Imax} for an unstabilized sample. The larger the Δt_{Imax} is, the more effective the addition of an antioxidant is in inhibiting oxidation. In Fig. 5c, E310 and BHT exhibited the highest stabilization ability with the Δt_{Imax} value of +25–30%. The largest negative stabilizing effect was found for Vitamin C, which was plausibly related to its high acidity [126, 127]. Many other antioxidants showed negligible effects on the recombinant spider silk lifetime, even though some of them are well-known antioxidants for other polymers, like Irganox 1098 for nylon 6. The reason why small molecules such as E310 and BHT were effective is that these molecules have the ability to

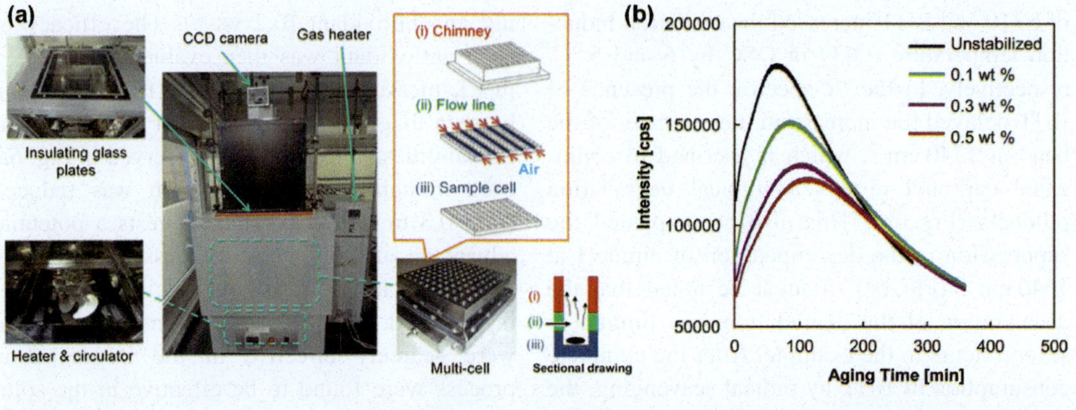

Fig. 5 **a** High-throughput chemiluminescence imaging instrument. **b** CL curves for recombinant spider silk powder stabilized by E310 recorded at 200 °C. Reprinted from [120] and [122] with permission from Elsevier

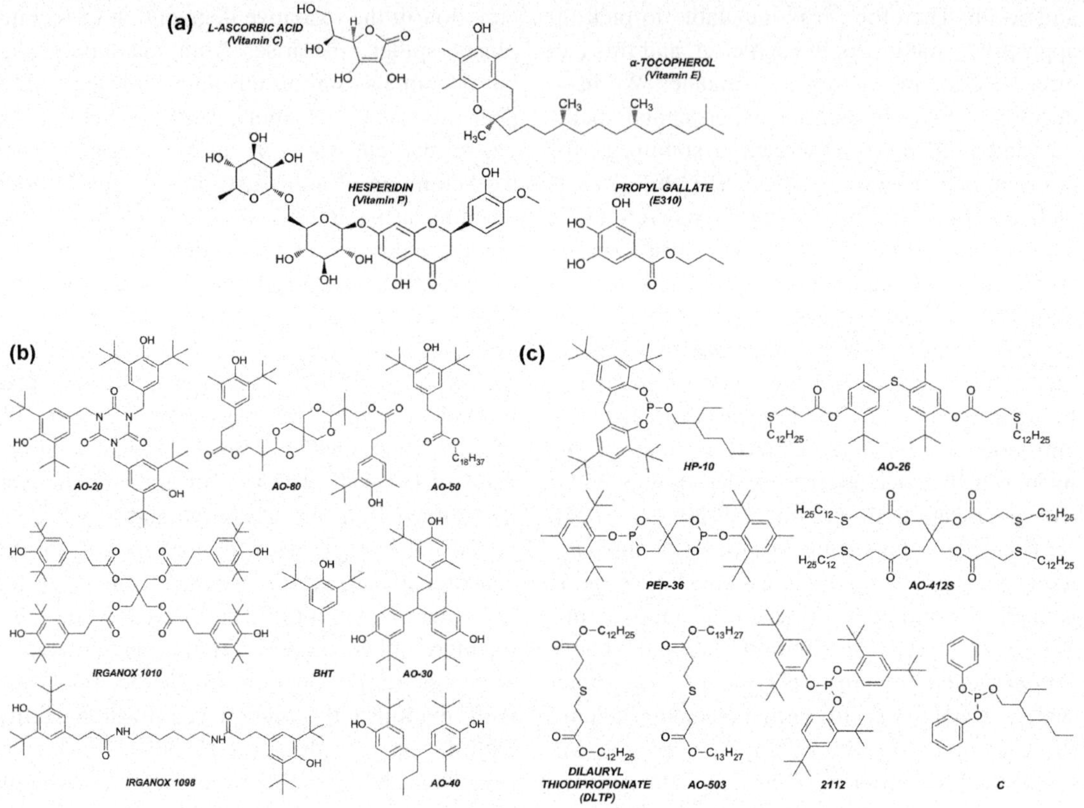

Scheme 1. Structure of tested antioxidants: **a** nature-derived antioxidants, **b** hindered phenols, **c** thioethers, and phosphites

penetrate the recombinant spider silk powder to some extent during impregnation, which is difficult for bulkier molecules. The effectiveness of E310 and BHT as stabilizers for recombinant spider silk powder was double-checked by DSC and IR measurements. The addition of 0.3 wt% of E310 and BHT increased the oxidation induction temperature (OIT) in DSC by 6 and 5 °C, respectively. In the IR spectra, the presence of BHT delayed the increase in the intensity of the band at 1740 cm^{-1}, which is ascribed to conjugated carbonyl groups as typical degradation products (Fig. 6b). This also accompanied the suppression in the decomposition of amide I at 1640 cm^{-1} (Fig. 6c). It must be noted that the deceleration of the degradation was limited to several hours in the example. After the complete consumption of BHT by radical scavenging, the degradation rate became similar to that for the unstabilized sample.

From the initial screening result in Fig. 6a, it appeared that the efficacy of antioxidants would be enhanced by improving the way of mixing antioxidants in the matrix. Thus, solution-cast films were prepared using a $CaCl_2/C_2H_5OH/H_2O$ solution of recombinant spider silk containing an antioxidant (0.2 wt%). The efficacy of each antioxidant was then evaluated by Δt_{Imax} in CL measurements at 200 °C (Fig. 6d). It can be seen that E310 and BHT kept high stabilization abilities similar to those observed in Fig. 6a, even though their concentration was reduced from 0.3 to 0.2 wt%. This suggests a potential advantage of the solution process in enhancing the stabilization efficacy due to improved mixing. Indeed, several phenolic antioxidants that were scarcely effective in the impregnation process were found to be effective in the solution process. In particular, Irganox 1098 exhibited the highest ability among all. Secondary

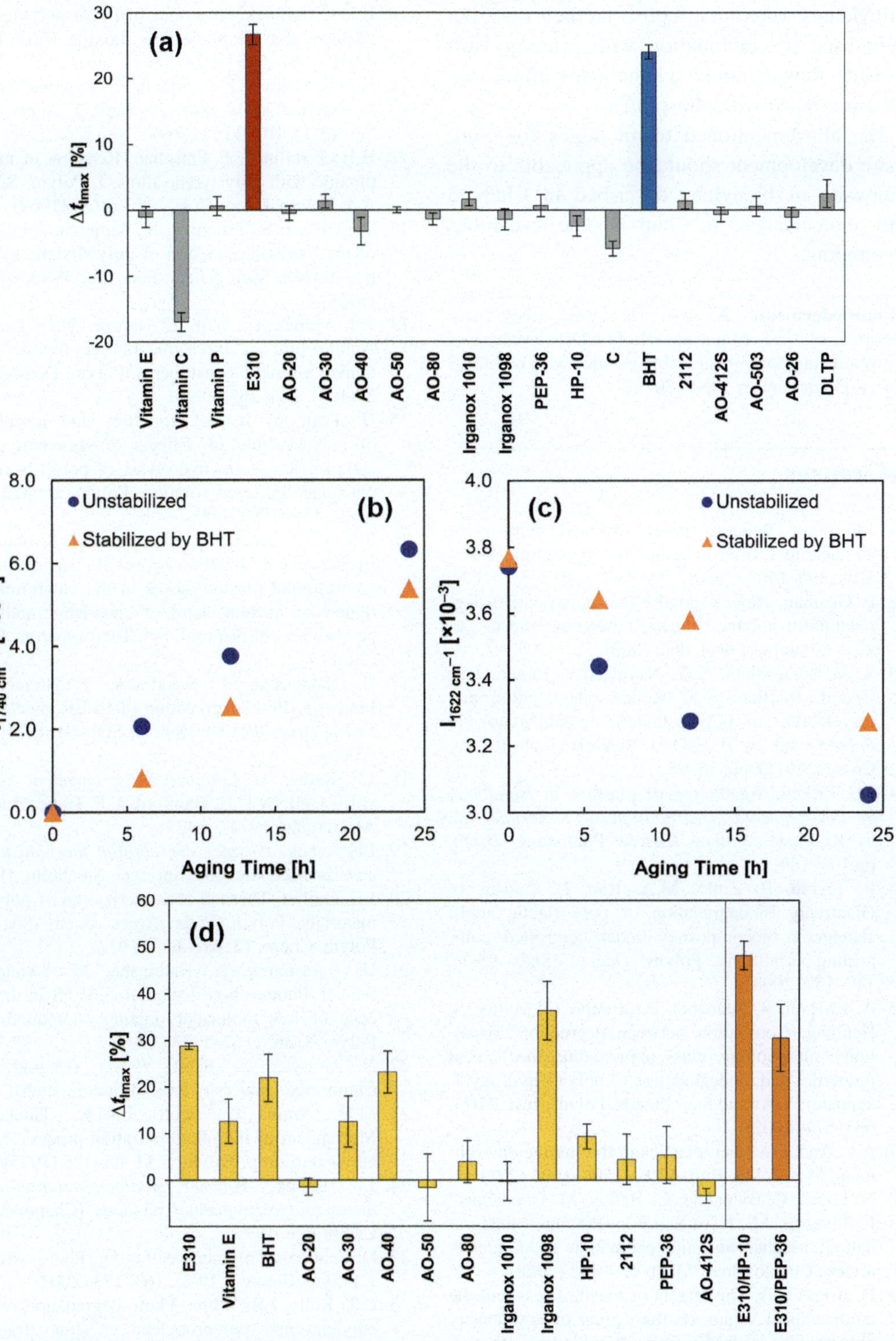

Fig. 6 **a** Screening results for 20 antioxidants at 200 °C. 0.3 wt% of antioxidants were added to spider silk powder by an impregnation method. The efficacy of the antioxidants was evaluated by Δt_{Imax}, which is the percentage change of t_{Imax} with respect to that for the unstabilized sample. **b**, **c** Development of the intensity of IR absorption bands at 1740 cm^{-1} and 1622 cm^{-1}. The former is attributed to conjugated carbonyl groups, while the latter to amide I. **d** Efficacy of antioxidants evaluated by Δt_{Imax} at 200 °C. Solution-cast films of recombinant spider silk containing 0.2 wt% of antioxidants were prepared. Reprinted from [122] with permission from Elsevier

antioxidants were not effective on their own, but when used in combination with primary antioxidants, they exhibited a synergistic effect, e.g. E310 combined with phosphites.

The abovementioned technologies for composite development should be applicable to the composite of bionylons described in Chap. 3 with photocatalysts in Chap. 4. The researches are ongoing.

Acknowledgements A part of this work was supported by JST Moonshot R&D "Development of photoswitching ocean-degradable plastics with edibility" Project (Grant number JPNP18016).

References

1. J.R. White, Polymer ageing: Physics, chemistry or engineering? Time to reflect. C. R. Chimie **9**(11–12), 1396–1408 (2006)
2. P. Gijsman, Review on the thermo-oxidative degradation of polymers during processing and in service. e-Polymers **8**(1), 065(2008)
3. S. Venkatachalam, S.G. Nayak, J.V. Labde, P.R. Gharal, K. Rao, A.K. Kelkar, Degradation and recyclability of poly (ethylene terephthalate). in *Polyester* ed. by H. E-D.M. Saleh (InTech, Rijeka, Croatia, 2012), pp. 75–98
4. J.E. Pickett, Weathering of plastics, in *Handbook of Environmental Degradation of Materials*. ed. by K. Myer (William Andrew Publishing, 2018), pp.163–184
5. R. Iovino, R. Zullo, M.A. Rao, L. Cassar, L. Gianfreda, Biodegradation of poly (lactic acid)/starch/coir biocomposites under controlled composting conditions. Polym. Degrad. Stab. **93**(1), 147–157 (2008)
6. A. El-Hadi, R. Schnabel, E. Straube, G. Müller, S. Henning, Correlation between degree of crystallinity, morphology, glass temperature, mechanical properties and biodegradation of poly (3-hydroxyalkanoate) PHAs and their blends. Polym. Test. **21**(6), 665–674 (2002)
7. A.L. Andrady, Microplastics in the marine environment. Mar. Pollut. Bull. **62**(8), 1596–1605 (2011)
8. N. Lucas, C. Bienaime, C. Belloy, M. Queneudec, F. Silvestre, J.E. Nava-Saucedo, Polymer biodegradation: Mechanisms and estimation techniques-A review. Chemosphere **73**(4), 429–442 (2008)
9. D. Briassoulis, The effects of tensile stress and the agrochemical Vapam on the ageing of low density polyethylene (LDPE) agricultural films. Part I. Mechanical behaviour. Polym. Degrad. Stab. **88**(3), 489–503 (2005)
10. H.H.G. Jellinek, Reactions of linear polymers with nitrogen dioxide and sulfur dioxide. Text. Res. J. **43**(10), 557–560 (1973)
11. H.H.G. Jellinek, F. Flajsman, F.J. Kryman, Reaction of SO_2 and NO_2 with polymers. J. Appl. Polym. Sci. **13**(1), 107–116 (1969)
12. H.H.G. Jellinek, F. Flajsman, Reaction of nitrogen dioxide with polystyrene films. J. Polym. Sci. Part A-1: Polym. Chem. **7**(4), 1153–1168 (1969)
13. T. Ogihara, S. Tsuchiya, K. Kuratani, Initial stage of the oxidation reaction of polyethylene by nitrogen dioxide. Bull. Chem. Soc. Jpn. **38**(6), 978–984 (1965)
14. R.J. Müller, U. Witt, E. Rantze, W.D. Deckwer, Architecture of biodegradable copolyesters containing aromatic constituents. Polym. Degrad. Stab. **59**(1–3), 203–208 (1998)
15. H. Tsuji, Y. Ikada, Properties and morphology of poly(l-lactide) 4. Effects of structural parameters on long-term hydrolysis of poly(l-lactide) in phosphate-buffered solution. Polym. Degrad. Stab. **67**(1), 179–189 (2000)
16. S.S. Ali, T. Elsamahy, E. Koutra, M. Kornaros, M. El-Sheekh, E. Abdelkarim, et al., Degradation of conventional plastic wastes in the environment. A review on current status of knowledge and future perspectives of disposal. Sci. Total Environ. 144719 (2021)
17. H. Nakamura, T. Nakamura, T. Noguchi, K. Imagawa, Photodegradation of PEEK sheets under tensile stress. Polym. Degrad. Stab. **91**(4), 740–746 (2006)
18. J.F. Rabek, in *Comprehensive chemical kinetics*, vol. 14, ed. by C.H. Bamford, C.F. Tipper (Elsevier, Amsterdam, 1974), p.425
19. J.F. Rabek, *Polymer degradation mechanisms and experimental methods* (Springer, Stocholm, 1994)
20. J.E. Guillet, Thermal characterization of polymeric materials. Polym. Prepr. Amer. Chem. Soc., Div. Polym. Chem. **13**, 628–632 (1972)
21. H. Kaczmarek, A. Kaminska, M. Swiatek, S. Sanyal, Photoinition degradation of PS in the presence of low molecular organic compounds. Eur. Polym. J. **36**, 1167–1173 (1999)
22. N. Grassie, G. Scott, *Polym* (Degrad. Stab. Cambridge University Press, London, 1985)
23. M.U. Amin, G. Scott, L.M.K. Tillekeratne, Mechanism of the photo initiation process in polyethylene. Europ. Polym. J. **11**, 85–126 (1975)
24. J.F. Rabek, *Polymer photodegradation-mechanisms and experimental methods* (Chapman Hall, Cambridge, 1995)
25. D. Feldman, Polymer weathering: Photo-oxidation. J. Polym. Environ. **10**(4), 163–173 (2002)
26. C.T. Kelly, J.R. White, Photo-degradation of polyethylene and polypropylene at slow strain-rate. Polym. Degrad. Stab. **56**(3), 367–383 (1997)
27. D.J. Carlsson, D.M. Wiles, The photodegradation of polypropylene films. II. Photolysis of ketonic

oxidation products. Macromolecules **2**(6), 587–597 (1969)
28. D.J. Carlsson, D.M. Wiles, The photodegradation of polypropylene films. III. Photolysis of polypropylene hydroperoxides. Macromolecules **2**(6), 597–606 (1969)
29. A. Roger, D. Sallet, J. Lemaire, Photochemistry of aliphatic polyamides. 4. Mechanisms of photooxidation of polyamides 6, 11, and 12 at long wavelengths. Macromolecules **19**(3), 579–584 (1986)
30. W.H. Sharkey, W.E. Mochel, Mechanism of the photoöxidation of amides. J. Am. Chem. Soc. **81**(12), 3000–3005 (1959)
31. H.M. Heuvel, K.C.J.B. Lind, ESR study of the initial reaction of the photodegradation of nylon 6. J. Polym. Sci. Part A-2: Polym. Phys. **8**(3), 401–410 (1970)
32. J.F. Rabek, *Mechanism of Photophysical Process and Photochemical Reaction in polymers* (John Wiley and Sons, New York, 1987)
33. H.J. Hueck, The biodeterioration of materials: An appraisal. Int. Biodeter. Biodegr. **48**, 5–11 (2001)
34. J.H. Walsh, Ecological considerations of biodeterioration. Int. Biodeter. Biodegr. **48**, 16–25 (2001)
35. S. Wallström, E. Strömberg, S. Karlsson, Microbiological growth testing of polymeric materials: an evaluation of new methods. Polym. Test. **24**(5), 557–563 (2005)
36. A. Lugauskas, L. Levinskaite, D. Peciulyte, Micromycetes as deterioration agents of polymeric materials. Int. Biodeter. Biodegr. **52**, 233–242 (2003)
37. E. Zanardini, P. Abbruscato, N. Ghedini, M. Realini, C. Sorlini, Influence of atmospheric pollutants on the biodeterioration of stone. Int. Biodeter. Biodegr. **45**, 35–42 (2000)
38. Y. Nuhoglu, E. Oguz, H. Uslu, A. Ozbek, B. Ipekoglu, I. Ocak et al., The accelerating effects of the microorganisms on biodeterioration of stone monuments under air pollution and continental-cold climatic conditions in Erzurum, Turkey. Sci. Total Environ. **364**, 272–283 (2006)
39. F. Faÿ, I. Linossier, J.-J. Peron, V. Langlois, K. Vallée-Rehel, Antifouling activity of marine paints: Study of erosion. Prog. Org. Coat. **60**, 194–206 (2007)
40. R.N. Tharanathan, Biodegradable films and composite coatings: Past, present and future. Trends Food Sci. Technol. **14**, 71–78 (2003)
41. J.D. Gu, Microbiological deterioration and degradation of synthetic polymeric materials: Recent research advances. Int. Biodeter. Biodegr. **52**, 69–91 (2003)
42. J.D. Gu, T. Ford, K. Thorp, R. Mitchell, Microbial growth on fiber reinforced composite materials. Int. Biodeter. Biodegr. **37**, 197–203 (1996)
43. H.-C. Flemming, Relevance of biofilms for the biodeterioration of surfaces of polymeric materials. Polym. Degrad. Stab. **59**, 309–315 (1998)
44. J.D. Gu, C. Lu, K. Thorp, A. Crasto, R. Mitchell, Fiber-reinforced polymeric composite are susceptible to microbial growth. J. Ind. Microbial. Biot. **18**, 364–369 (1997)
45. J.D. Gu, M. Roman, T. Esselman, R. Mitchell, The role of microbial biofilms in deterioration of space station candidate materials. Int. Biodeter. Biodegr. **41**, 25–33 (1998)
46. F. Capitelli, P. Principi, C. Sorlini, Biodeterioration of modern materials in contemporary collections: can biotechnology help? Trends Biotechnol. **24**(8), 350–354 (2006)
47. S. Bonhomme, A. Cuer, A.-M. Delort, J. Lemaire, M. Sancelme, G. Scott, Environmental biodegradation of polyethylene. Polym. Degrad. Stab. **81**, 441–452 (2003)
48. A. Banerjee, K. Chatterjee, G. Madras, Enzymatic degradation of polymers: A brief review. Mater. Sci. Technol. **30**(5), 567–573 (2014)
49. J.D. Gu, T.E. Ford, D.B. Mitton, R. Mitchell, Microbial degradation and deterioration of polymeric materials. in *The Uhlig corrosion handbook*, 2nd edn, ed. by W. Revie (Wiley, New York, 2000), pp. 439–460
50. H. Vieyra, M.A. Aguilar-Méndez, E. San Martín-Martínez, Study of biodegradation evolution during composting of polyethylene-starch blends using scanning electron microscopy. J. Appl. Polym. Sci. **127**(2), 845–853 (2013)
51. J.M. Raquez, A. Bourgeois, H. Jacobs, P. Degée, M. Alexandre, P. Dubois, Oxidative degradations of oxodegradable LDPE enhanced with thermoplastic pea starch: Thermo-mechanical properties, morphology, and UV-ageing studies. J. Appl. Polym. Sci. **122**(1), 489–496 (2011)
52. F. Amri, S. Husseinsyah, K. Hussin, Mechanical, morphological and thermal properties of chitosan filled polypropylene composites: The effect of binary modifying agents. Compos. -A: Appl. Sci. Manuf. **46**, 89–95 (2013)
53. M. Sunilkumar, T. Francis, E.T. Thachil, A. Sujith, Low density polyethylene-chitosan composites: A study based on biodegradation. Chem. Eng. J. **204**, 114–124 (2012)
54. T. Lundin, S.M. Cramer, R.H. Falk, C. Felton, Accelerated weathering of natural fiber-filled polyethylene composites. J. Mater. Civ. Eng. **16**(6), 547–555 (2004)
55. D. Ołdak, H. Kaczmarek, T. Buffeteau, C. Sourisseau, Photo-and bio-degradation processes in polyethylene, cellulose and their blends studied by ATR-FTIR and Raman spectroscopies. J. Mater. Sci. **40**(16), 4189–4198 (2005)
56. A. Kausar, Scientific potential of chitosan blending with different polymeric materials: A review. J. Plast. Film Sheeting **33**(4), 384–412 (2017)
57. F. Croisier, C. Jérôme, Chitosan-based biomaterials for tissue engineering. Eur. Polym. J. **49**(4), 780–792 (2013)

58. D.R. Barleany, Emerging trends in the development of chitosan films for active food packaging. World Chem. Eng. J. **4**(2), 46–50 (2020)
59. Y.G. Thyavihalli-Girijappa, S. Mavinkere-Rangappa, J. Parameswaranpillai, S. Siengchin, Natural fibers as sustainable and renewable resource for development of eco-friendly composites: A comprehensive review. Front. Mater. **6**, 226 (2019)
60. A. Rezaei Kolahchi, A. Ajji, P.J. Carreau, Enhancing hydrophilicity of polyethylene terephthalate surface through melt blending. Polym. Engin. Sci. **55**(2), 349–358 (2015)
61. Y. Feng, G. Han, T.S. Chung, M. Weber, N. Widjojo, C. Maletzko, Effects of polyethylene glycol on membrane formation and properties of hydrophilic sulfonated polyphenylenesulfone (sPPSU) membranes. J. Membr. Sci. **531**, 27–35 (2017)
62. J. Natarajan, G. Madras, K. Chatterjee, Development of graphene oxide/galactitol polyester-based biodegradable composites for biomedical applications. ACS Omega **2**(9), 5545–5556 (2017)
63. F. Wang, Y. Wu, Y. Huang, Novel application of graphene oxide to improve hydrophilicity and mechanical strength of aramid nanofiber hybrid membrane. Compos. Part A Appl. Sci. **110**, 126–132 (2018)
64. D. Ribitsch, E. Herrero Acero, A. Przylucka, S. Zitzenbacher, A. Marold, C. Gamerith, et al., Enhanced cutinase-catalyzed hydrolysis of polyethylene terephthalate by covalent fusion to hydrophobins. Appl. Environ. Microbiol. **81**(11), 3586–3592 (2015)
65. N. Shimpi, M. Borane, S. Mishra, M. Kadam, Biodegradation of polystyrene (PS)-poly (lactic acid) (PLA) nanocomposites using Pseudomonas aeruginosa. Macromol. Res. **20**(2), 181–187 (2012)
66. I. Cacciari, P. Quatrini, G. Zirletta, E. Mincione, V. Vinciguerra, P. Lupattelli et al., Isotactic polypropylene biodegradation by a microbial community: physicochemical characterization of metabolites produced. Appl. Environ. Microbiol. **59**(11), 3695–3700 (1993)
67. J. Shang, M. Chai, Y. Zhu, Solid-phase photocatalytic degradation of polystyrene plastic with TiO_2 as photocatalyst. J. Solid State Chem. **174**(1), 104–110 (2003)
68. Y. Lei, H. Lei, J. Huo, Innovative controllable photocatalytic degradation of polystyrene with hindered amine modified aromatic polyamide dendrimer/polystyrene-grafted-TiO_2 photocatalyst under solar light irradiation. Polym. Degrad. Stab. **118**, 1–9 (2015)
69. S.B. Ameur, A. Barhoumi, B. Duponchel, G. Leroy, M. Amlouk, H. Guermazi, Physical investigations and photocatalytic activities on ZnO and SnO_2 thin films deposited on flexible polymer substrate. Vacuum **155**, 546–552 (2018)
70. T.S. Tofa, K.L. Kunjali, S. Paul, J. Dutta, Visible light photocatalytic degradation of microplastic residues with zinc oxide nanorods. Environ. Chem. Lett. **17**(3), 1341–1346 (2019)
71. G.C. De Assis, E. Skovroinski, V.D. Leite, M.O. Rodrigues, A. Galembeck, M.C. Alves et al., Conversion of "Waste Plastic" into photocatalytic nanofoams for environmental remediation. ACS Appl. Mater. Interfaces **10**(9), 8077–8085 (2018)
72. T. Uekert, H. Kasap, E. Reisner, Photoreforming of nonrecyclable plastic waste over a carbon nitride/nickel phosphide catalyst. J. Am. Chem. Soc. **141**(38), 15201–15210 (2019)
73. T. Uekert, M.F. Kuehnel, D.W. Wakerley, E. Reisner, Plastic waste as a feedstock for solar-driven H_2 generation. Ener. Environ. Sci. **11**(10), 2853–2857 (2018)
74. K. Kalyanasundaram, M. Grätzel, Applications of functionalized transition metal complexes in photonic and optoelectronic devices. Coord. Chem. Rev. **177**(1), 347–414 (1998)
75. S. Hamad, C.R.A. Catlow, S.M. Woodley, S. Lago, J.A. Mejías, Structure and stability of small TiO_2 nanoparticles. J. Phys. Chem. B **109**(33), 15741–15748 (2005)
76. M. Adachi, Y. Murata, J. Takao, J. Jiu, M. Sakamoto, F. Wang, Highly efficient dye-sensitized solar cells with a titania thin-film electrode composed of a network structure of single-crystal-like TiO_2 nanowires made by the "oriented attachment" mechanism. J. Am. Chem. Soc. **126**(45), 14943–14949 (2004)
77. O.K. Varghese, M. Paulose, T.J. LaTempa, C.A. Grimes, High-rate solar photocatalytic conversion of CO_2 and water vapor to hydrocarbon fuels. Nano Lett. **9**(2), 731–737 (2009)
78. K. Zakrzewska, Mixed oxides as gas sensors. Thin Solid Films **391**(2), 229–238 (2001)
79. M. Tu-morn, N. Pairoh, W. Sutapun, T. Trongsatitkul, Effects of titanium dioxide nanoparticle on enhancing degradation of polylactic acid/low density polyethylene blend films. Mater. Today: Proc. **17**, 2048–2061 (2019)
80. M.A. Sani, M. Maleki, H. Eghbaljoo-Gharehgheshlaghi, A. Khezerlou, E. Mohammadian, Q. Liu et al., Titanium dioxide nanoparticles as multifunctional surface-active materials for smart/active nanocomposite packaging films. Adv. Colloid Interface Sci.102593 (2021)
81. W. Liang, Y. Luo, S. Song, X. Dong, X. Yu, High photocatalytic degradation activity of polyethylene containing polyacrylamide grafted TiO_2. Polym. Degrad. Stab. **98**(9), 1754–1761 (2013)
82. L. Zan, W. Fa, S. Wang, Novel photodegradable low-density polyethylene-TiO_2 Nanocomposite film. Environ. Sci. Technol. **40**(5), 1681–1685 (2006)
83. Y. Zhang, T. Sun, D. Zhang, Z. Shi, X. Zhang, C. Li et al., Enhanced photodegradability of PVC plastics

film by codoping nano-graphite and TiO_2. Polym. Degrad. Stab. **181**, 109332 (2020)
84. H. Dong, G. Zeng, L. Tang, C. Fan, C. Zhang, X. He, Y. He, An overview on limitations of TiO_2-based particles for photocatalytic degradation of organic pollutants and the corresponding countermeasures. Water Res. **79**, 128–146 (2015)
85. M. Anpo, Use of visible light. Second-generation titanium oxide photocatalysts prepared by the application of an advanced metal ion-implantation method. Pure Appl. Chem. **72**(9), 1787–1792 (2000)
86. I. Bannat, K. Wessels, T. Oekermann, J. Rathousky, D. Bahnemann, M. Wark, Improving the photocatalytic performance of mesoporous titania films by modification with gold nanostructures. Chem. Mater. **21**(8), 1645–1653 (2009)
87. M. Khairy, W. Zakaria, Effect of metal-doping of TiO_2 nanoparticles on their photocatalytic activities toward removal of organic dyes. Egypt. J. Pet **23**(4), 419–426 (2014)
88. D. Chen, Z. Jiang, J. Geng, Q. Wang, D. Yang, Carbon and nitrogen co-doped TiO_2 with enhanced visible-light photocatalytic activity. Ind. Eng. Chem. Res. **46**(9), 2741–2746 (2007)
89. F. Dong, S. Guo, H. Wang, X. Li, Z. Wu, Enhancement of the visible light photocatalytic activity of C-doped TiO_2 nanomaterials prepared by a green synthetic approach. J. Phys. Chem. C **115**(27), 13285–13292 (2011)
90. J. Zhu, W. Zheng, B. He, J. Zhang, M. Anpo, Characterization of Fe-TiO_2 photocatalysts synthesized by hydrothermal method and their photocatalytic reactivity for photodegradation of XRG dye diluted in water. J. Mol. Catal. Chem. **216**(1), 35–43 (2004)
91. P.D. Cozzoli, A. Kornowski, H. Weller, Low-temperature synthesis of soluble and processable organic-capped anatase TiO_2 nanorods. J. Am. Chem. Soc. **125**(47), 14539–14548 (2003)
92. G. Hasegawa, K. Kanamori, K. Nakanishi, T. Hanada, Facile preparation of transparent monolithic titania gels utilizing a chelating ligand and mineral salts. J. Solgel Sci. Technol. **53**(1), 59–66 (2010)
93. M. Kakihana, M. Kobayashi, K. Tomita, V. Petrykin, Application of water-soluble titanium complexes as precursors for synthesis of titanium-containing oxides via aqueous solution processes. Bull. Chem. Soc. Jpn. **83**(11), 1285–1308 (2010)
94. L. Wei, C. Yu, Q. Zhang, H. Liu, Y. Wang, TiO_2-based heterojunction photocatalysts for photocatalytic reduction of CO_2 into solar fuels. J. Mater. Chem. A **6**(45), 22411–22436 (2018)
95. Y. Bessekhouad, D. Robert, J.V. Weber, Bi_2S_3/TiO_2 and CdS/TiO_2 heterojunctions as an available configuration for photocatalytic degradation of organic pollutant. J. Photochem. Photobiol. Chem. **163**(3), 569–580 (2004)
96. G. Lui, J.Y. Liao, A. Duan, Z. Zhang, M. Fowler, A. Yu, Graphene-wrapped hierarchical TiO_2 nanoflower composites with enhanced photocatalytic performance. J. Mater. Chem. **1**(39), 12255–12262 (2013)
97. K. Zhou, Y. Zhu, X. Yang, X. Jiang, C. Li, Preparation of graphene-TiO_2 composites with enhanced photocatalytic activity. New J. Chem. **35**(2), 353–359 (2011)
98. L. Vicarelli, S.J. Heerema, C. Dekker, H.W. Zandbergen, Controlling defects in graphene for optimizing the electrical properties of graphene nanodevices. ACS Nano **9**(4), 3428–3435 (2015)
99. J. Haskins, A. Kınacı, C. Sevik, H. Sevinçli, G. Cuniberti, T. Çağın, Control of thermal and electronic transport in defect-engineered graphene nanoribbons. ACS Nano **5**(5), 3779–3787 (2011)
100. X. Liu, L. Pan, T. Lv, G. Zhu, T. Lu, Z. Sun, C. Sun, Microwave-assisted synthesis of TiO_2-reduced graphene oxide composites for the photocatalytic reduction of Cr(VI). RSC Adv. **1**(7), 1245–1249 (2011)
101. R. Larciprete, S. Fabris, T. Sun, P. Lacovig, A. Baraldi, S. Lizzit, Dual path mechanism in the thermal reduction of graphene oxide. J. Am. Chem. Soc. **133**(43), 17315–17321 (2011)
102. T. Xu, L. Zhang, H. Cheng, Y. Zhu, Significantly enhanced photocatalytic performance of ZnO via graphene hybridization and the mechanism study. Appl. Catal. B-Environ. **101**(3), 382–387 (2011)
103. R. Sellappan, J. Sun, A. Galeckas, N. Lindvall, A. Yurgens, A.Y. Kuznetsov et al., Influence of graphene synthesizing techniques on the photocatalytic performance of graphene-TiO_2 nanocomposites. Phys. Chem. Chem. Phys. **15**(37), 15528–15537 (2013)
104. G. Jiang, Z. Lin, C. Chen, L. Zhu, Q. Chang, N. Wang et al., TiO_2 nanoparticles assembled on graphene oxide nanosheets with high photocatalytic activity for removal of pollutants. Carbon **49**(8), 2693–2701 (2011)
105. N.N.T. Ton, A.T.N. Dao, K. Kato, T. Ikenaga, D.X. Trinh, T. Taniike, One-pot synthesis of TiO_2/graphene nanocomposites for excellent visible light photocatalysis based on chemical exfoliation method. Carbon **133**, 109–117 (2018)
106. M. Cai, D. Thorpe, D.H. Adamson, H.C. Schniepp, Methods of graphite exfoliation. J. Mater. Chem. **22**(48), 24992–25002 (2012)
107. R. Narayan, S.O. Kim, Surfactant mediated liquid phase exfoliation of graphene. Nano Converg. **2**(1), 1–19 (2015)
108. N.N.T. Ton, M.Q. Ha, T. Ikenaga, A. Thakur, H.C. Dam, T. Taniike, Solvent screening for efficient chemical exfoliation of graphite. 2D Materials **8**(1), 015019 (2020)
109. M.A. Pimenta, G. Dresselhaus, M.S. Dresselhaus, L.G. Cancado, A. Jorio, R. Saito, Studying disorder

in graphite-based systems by Raman spectroscopy. Phys. Chem. Chem. Phys. **9**(11), 1276–1290 (2007)
110. W. Zhang, J. Cui, C.A. Tao, Y. Wu, Z. Li, L. Ma et al., A strategy for producing pure single-layer graphene sheets based on a confined self-assembly approach. Angew. Chem. Int. Ed. **121**(32), 5978–5982 (2009)
111. Y. Zhang, N. Zhang, Z.R. Tang, Y.J. Xu, Improving the photocatalytic performance of graphene-TiO_2 nanocomposites via a combined strategy of decreasing defects of graphene and increasing interfacial contact. Phys. Chem. Chem. Phys. **14**(25), 9167–9175 (2012)
112. A.C. Ferrari, J.C. Meyer, V. Scardaci, C. Casiraghi, M. Lazzeri, F. Mauri et al., Raman spectrum of graphene and graphene layers. Phys. Rev. Lett. **97**(18), 187401 (2006)
113. F. Tuinstra, J.L. Koenig, Raman spectrum of graphite. J. Chem. Phys. **53**(3), 1126–1130 (1970)
114. K.N. Kudin, B. Ozbas, H.C. Schniepp, R.K. Prud'Homme, I.A. Aksay, R. Car, Raman spectra of graphite oxide and functionalized graphene sheets. Nano Lett. **8**(1), 36–41 (2008)
115. K.S. Rao, J. Senthilnathan, Y.F. Liu, M. Yoshimura, Role of peroxide ions in formation of graphene nanosheets by electrochemical exfoliation of graphite. Sci. Rep. **4**, 4237 (2014)
116. P. Yu, S.E. Lowe, G.P. Simon, Y.L. Zhong, Electrochemical exfoliation of graphite and production of functional graphene. Curr. Opin. Colloid Interface Sci. **20**(5), 329–338 (2015)
117. Y. Yang, E. Liu, H. Dai, L. Kang, H. Wu, J. Fan et al., Photocatalytic activity of Ag-TiO_2-graphene ternary nanocomposites and application in hydrogen evolution by water splitting. Int. J. Hydrogen Energy **39**(15), 7664–7671 (2014)
118. Y. Hao, Y. Wang, L. Wang, Z. Ni, Z. Wang, R. Wang et al., Probing layer number and stacking order of few-layer graphene by Raman spectroscopy. Small **6**(2), 195–200 (2010)
119. F. Bonaccorso, A. Lombardo, T. Hasan, Z. Sun, L. Colombo, A.C. Ferrari, Production and processing of graphene and 2d crystals. Mater. Today **15**(12), 564–589 (2012)
120. N. Aratani, I. Katada, K. Nakayama, M. Terano, T. Taniike, Development of high-throughput chemiluminescence imaging instrument for parallel evaluation of polymer lifetime. Polym. Degrad. Stab. **121**, 340–347 (2015)
121. T. Taniike, T. Kitamura, K. Nakayama, K. Takimoto, N. Aratani, T. Wada et al., Stabilizer formulation based on high-throughput chemiluminescence imaging and machine learning. ACS Appl. Polym. Mater. **2**(8), 3319–3326 (2020)
122. A.T.N. Dao, J.I. Shimokata, K. Takeuchi, K. Nakayama, T. Taniike, Stabilization of recombinant spider silk in thermo-oxidative degradation: High-throughput screening for antioxidants. Polym. Degrad. Stab. **153**, 37–46 (2018)
123. A.T.M. Mai, A. Thakur, N.N.T. Ton, T.N. Nguyen, T. Kaneko, T. Taniike, Photodegradation of a semi-aromatic bio-derived polyimide. Polym. Degrad. Stab. **184**, 109472 (2021)
124. I. Blakey, G.A. George, N.C. Billingham, Energy transfer and activated chemiluminescence during thermal oxidation of polypropylene: Evidence for chemically induced electron exchange luminescence. Macromolecules **34**(26), 9130–9138 (2001)
125. K. Nakayama, A.T. Dao, K. Takeuchi, J.I. Shimokata, T. Taniike, Chemiluminescence from recombinant spider silk under thermo-oxidative degradation. Polym. Degrad. Stab. **144**, 264–269 (2017)
126. A. Tsugita, J. Scheffler, A rapid method for acid hydrolysis of protein with a mixture of trifluoroacetic acid and hydrochloric acid. Eur. J. Biochem. **124**, 585–588 (1982)
127. M. Fountoulakis, H.-W. Lahm, Hydrolysis and amino acid composition analysis of proteins. J. Chromatogr. A **826**, 109–134 (1998)

Biodegradation of Biodegradable Plastics in Seawater

Atsuyoshi Nakayama

1 Marine Plastic Litter Problem and Biodegradable Plastics

The production of plastics has been increasing and now the global production of plastics is over 400 million tons, of which 70–80% is discharged as waste plastic. Some of them are dispersed into the environment and eventually end up in the ocean through rivers. It is said that this amount exceeds 10 million tons every year. Marine plastic debris is thought to be degraded and fragmented into microplastics. It has been pointed out that microplastics absorb and concentrate very small amounts of toxic organic substances that are widely spread in seawater and the toxic substances enter the ecosystem through oral ingestion of the microplastics by fish and other organisms. Microplastics should not exist in a sustainable society.

Biodegradable plastics are expected to play a role in solving the microplastic problem by deploying in applications where marine litter is likely occurring.

The biodegradable plastics attracted attention in the 1990s from the viewpoint of extending the life of final landfill disposal sites and preventing littering, and several resins were developed and their biodegradability was studied, but attention was mainly focused on the biodegradation on land and composting. There are not many research reports on biodegradation in the ocean, which means that there is not enough knowledge about whether commonly known biodegradable plastics biodegrade in the ocean or not. There are cautious opinions about the effectiveness and cost of biodegradation in the ocean, as well as concerns that the introduction of biodegradable plastics will conversely increase plastic waste in the environment [1]. Table 1 shows the list of standardization of biodegradation test methods. Many standard test methods developed around 2000 were related to terrestrial biodegradation or biogasification, and the method for marine biodegradation was developed only in 2016. Since then, a series of standard test methods in the marine environment have been developed.

Furthermore, studies on certification systems, confirmation of marine biodegradability of existing biodegradable plastics, and development of new materials are underway.

2 Biodegradable Plastics

Biodegradable plastics are resins that are completely metabolized to carbon dioxide and water by microorganisms in the environment, and they attracted much attention around 1990.

A. Nakayama (✉)
National Institute of Advanced Industrial Science and Technology, AIST, 1-8-31 Midorigaoka, Ikeda 563-8577, Osaka, Japan
e-mail: a.nakayama@aist.go.jp

T. Kaneko (ed.), *Photo-switched Biodegradation of Bioplastics in Marine Environments*,
https://doi.org/10.1007/978-981-99-4354-8_7

Table 1 ISO for aerobic biodegradation test methods

BO	
14851	Determination of the ultimate aerobic biodegradability of plastic materials in an aqueous medium—Method by measuring the oxygen dem and in a closed respirometer
14852	Determination of the ultimate aerobic biodegradability of plastic materials in an aqueous medium—Method by analysis of evolved carbon di oxide
14855	Determination of the ultimate aerobic biodegradability of plastic materials under controlled composting conditions—Method by analysis of evolved carbon dioxide—Part 2: Gravimetric measurement of carbon di oxide evolved in a laboratory-scale test
16929	Plastics—Determination of the degree of disintegration of plastic materials under defined com posting conditions in a pilot-scale test
17556	Plastics—Determination of the ultimate aerobic biodegradability of plastic materials in soil by measuring the oxygen dem and in a respirometer or the amount of carbon di oxide evolved
18830	Plastics—Determination of aerobic biodegradation of non-floating plastic materials in a seawater/sandy sediment interface—Method by measuring the oxygen demand in closed respirometer
19679	Plastics—Determination of aerobic biodegradation of non-floating plastic materials in a seawater/sediment interface—Method by analysis of evoled carbon di oxide
20200	Plastics—Determination of the degree of disintegration of plastic materials under simulated com posting conditions in a laboratory-scale test
22766	Plastics—Determination of the degree of disintegration of plastic materials in marine habitats under real field conditions
23977	Plastics—Determination of the aerobic biodegradation of plastic materials exposed to seawater—Part 1: Method by analysis of evolved carbon d oxide

The first biodegradable plastic to be put to practical use was poly(3-hydroxybutyrate-co-3-hydroxyvalerate), (PHBV), which is biosynthesized by microorganisms in bacteria and used for high-end cosmetic containers. Poly(ε-caprolactone), (PCL), poly(butylene succinate) and its copolymers (PBS and PBSA), PBS/carbonate, polyethylene succinate (PES), and aromatic-containing plastics (PBAT, PBST) copolymerized with terephthalic acid in certain proportions were also presented. Several plastic manufacturers announced and launched polylactic acid (PLA). Various types of polyester-based plastics have been developed, but biosynthetic plastics have problems such as thermal decomposition and odor during molding, while chemosynthetic polymers also have problems in heat resistance and mechanical strength. These problems can be overcome by the introduction of aromatic rings but at the expense of biodegradability. Other biodegradable plastics presented include polyvinyl alcohol (PVA) and natural product-based modified starch and cellulose acetate. In the decade since 1990s, the plastic waste that littered the environment has become an issue, and it had been pointed out that bulky plastics shorten the life of final landfill disposal facilities. As a result, biodegradable plastics have attracted attention, but they have not been widely used except for some agriculture, forestry, fisheries and civil engineering fields because of cost, thermal and mechanical properties.

In the 2000s, with the rise of the global warming problem, interest shifted to biomass-derived plastics and biobased polymers, although they are in the same category as bioplastics. Biodegradable plastics are positioned as environmentally low-impact materials to deal with the environmental pollution problem caused by dissipated plastics, whereas biobased materials are made from biomass instead of fossil resources, aiming at environmentally low-impact plastics that do not increase GHGs and help to control global warming. The sharp rise in crude oil prices since 2004 has been a tailwind for biobased materials, and production of PET, PE, and PP from biomass raw materials has begun, and new biobased materials such as polytrimethylene terephthalate (PTT) and polyethylene furanoate (PEF) have been developed.

In the 2010s, the existence of microplastics in the ocean and the impact of waste plastics on the ecosystem, as symbolized by the sea turtle video, came under close scrutiny, and biodegradable plastics and marine biodegradable plastics again attracted attention as one of the possible solutions to the marine plastic problem. Among the factors that prevented the spread of biodegradable plastics in the past, the technical aspects have been improved, and the environment for the spread of biodegradable plastics has improved, as the cost of biodegradable plastics has been reduced and the low environmental impact of existing plastics has become more important, such as regulations on existing plastics. The chemical structures are basically similar to those used in the past, with biosynthetic plastics such as PHBH, PHBV, and P3HB4HB, and chemically synthesized plastics such as PBS, PBSA, PBAT, PESe, PETS, polyglycolic acid, and starch polyester becoming more widespread.

Compared to microbial polymers, synthetic polyesters have various main chain structures, and their biodegradability cannot be generalized. Ester hydrolysis enzymes such as lipase play a key role in the biodegradation of polyester main chain, which is the rate-limiting step, and side chain substituents and aromatic structures inhibit the biodegradation. The longer the repeating unit, the smaller the ester content in the unit becomes and the more hydrophobic it becomes, the less biodegradable it becomes. Conversely, the shorter the repeating unit, the more rigid the main chain becomes and the less biodegradable it becomes. Hydrophobic groups have the effect of suppression of biodegradation, and even hydrophobic groups at the ends alone have a negative effect [2]. Copolymerization has a great effect to promote biodegradability. Although biodegradation of polylactic acid is slow in natural environment, copolymerization of L-lactide with 10–20% of ε-caprolactone can greatly improve the biodegradability [3]. Copolymerization is also used to improve mechanical strength by introducing terephthalic acid and amide units, but biodegradability is reduced when the introduction ratio is increased [4]. Urethanation with diisocyanate compounds is an effective method for chain extension, but biodegradability varies greatly depending on the type of isocyanate compound. That is, aromatic or bulky isocyanates suppress biodegradability. The amino-acid-derived lysine diisocyanate system showed good biodegradability, whereas the biodegradability of 4,4'-methylene diphenyl diisocyanate (MDI) and isophorone diisocyanate (IPDI) was considerably suppressed. Biodegradability of plastics derived from polysaccharides is greatly affected by the structure and degree of substitution of substituents introduced for thermoplasticization. Polyamide 4 is a type of nylon composed of repeating unit of γ-aminobutyric acid, which is also positioned as a polyamino acid, and is categorized to be an of engineering plastic having good biodegradation [5]. Although there are many reports on the biodegradation of these plastics by activated sludge and compost, there are still few reports on marine biodegradation of these plastics. Unlike plastics located on land, plastics distributed in the ocean diffuse easily over a wide area, so it is necessary to accumulate more data on the biodegradability and safety of biodegradable plastics in the ocean.

3 Evaluation Method of Marine Biodegradation

There are two methods for measuring marine biodegradation: laboratory tests and field tests (Table 2). In laboratory tests, natural seawater or seawater mixed with seabed sediment is commonly used as a microbial source in the established standard test methods of ASTM D 7081, OECD 306, ISO 18830 and ISO 19679. In this method, seawater mixed with seabed soil (sediment) is used as an inoculum source and tested under static conditions, which is a biodegradation test method in an environment that simulates

Table 2 Comparison between controlled laboratory biodegradation test and disintegration test in the ocean

	Laboratory test	Field test
	(Biodegradation test)	(Disintegration test)
Measurement	Comsumed O_2	Change of weight
	or evolved CO_2	
Degrader	Microorganism	Microorganism
		Other marine livings
		Physical and chemical stress
Points to note	Possibility of part of the carbon in the resin being taken up by the bacteria	Possibility of breaking into pieces and falling off

the settling of resin on the seabed. The authors have investigated marine biodegradation of various polymers using a method based on ISO 14851 and ISO 14852, which is a test method assuming a floating state in which surface natural seawater itself is used instead of an inorganic nutrient medium and activated sludge (Fig. 1). In either method, the amount of carbon dioxide produced or oxygen consumed by biodegradation is evaluated (Fig. 2), since the resin is eventually converted to carbon dioxide through biodegradation. Artificial seawater can be used instead of natural seawater, but in this case, it is necessary to add a separate microbial source, for example, a consortium of 10 microorganisms from ASTM D 6691. Laboratory evaluation methods for marine biodegradation have been further developed as ISO 23977, and new methods are

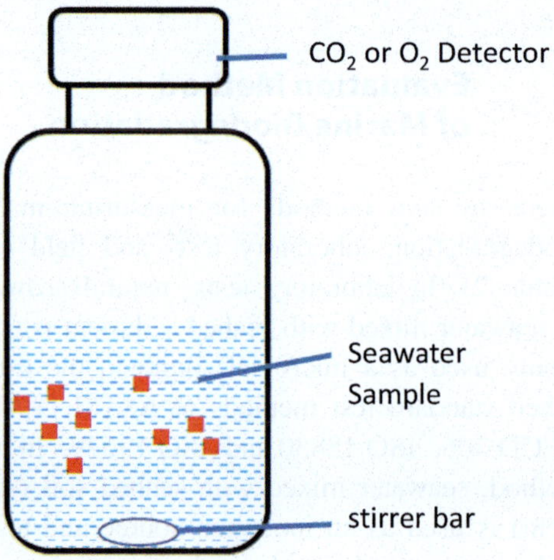

Fig. 1 Aerobic biodegradation test apparatus

being considered. The field test is also important as an evaluation method of marine biodegradability of plastics. The field test is carried out by immersing the plastic sample in the real sea and measuring the weight loss of the sample, that is, the field test is not a biodegradation test but a disintegration test. In principle, the weight loss of biodegradable plastic of the field test is proportional to the carbon dioxide production or the oxygen consumption of the laboratory test, but in reality, it often the case that the biodegradation proceeds slower for the laboratory test because the decomposed plastic is partly used as carbon source to constitute the microbial fungus and not detected by carbon dioxide nor oxygen consumption. Therefore, in the laboratory test, even if the resin biodegrades and disappears, there is a time lag for complete is conversion to carbon dioxide until self-digestion of the microbes. In addition to these differences based on the measurement principle, the following differences in the experimental environment are also significant. First, laboratory tests have shown that zooplankton or crustacean larvae quickly die in test flasks, and only microbes are responsible for biodegradation. In the real ocean, on the other hand, these planktons and a variety of other marine organisms can be involved in the biodegradation process. In the laboratory test, only the bacteria existing in the several hundred milliliters of seawater sampled are involved, but in the real environment, the microbes are in contact with new seawater continuously. We have just started to examine how these differences affect the results of the laboratory tests and the field tests.

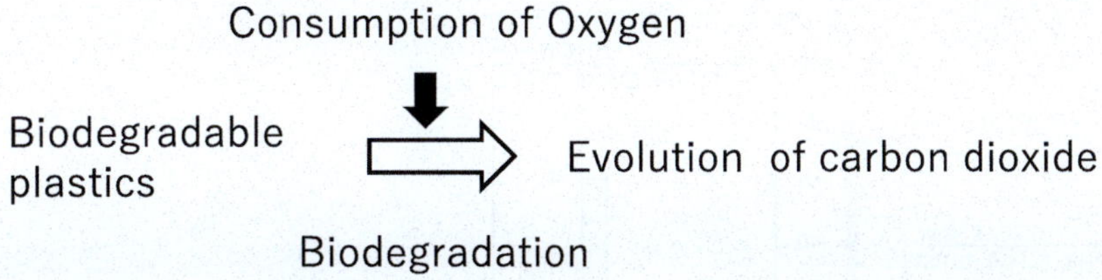

Fig. 2 Biodegradation and mineralization of plastics

4 Seawater Biodegradability of Biodegradable Plastics

Figure 3 shows the biodegradation rates of various plastics after four weeks of laboratory biodegradation tests using natural seawater collected from various locations and seawater collected in Osaka (urban) on the same day. Three typical biodegradable plastics, microbial, synthetic, and nylon, were used. The enzymes and bacterial flora that work in the biodegradation process are different among these plastics. In all seawaters, P3HB showed the fastest biodegradation [6–8]. The biodegradation activity of seawater varies greatly depending on the location where the water is sampled. For example, the biodegradation activity of seawater from Yokohama (urban) was similar to that of Osaka, whereas that of seawater from Muroto and Kagoshima was much lower, about one-fifth of that of Osaka. The biodegradation activity tended to be higher in seawater-facing urban areas with a high population concentration, such as Osaka and Yokohama, which is considered to be due to the effect of water quality. Microorganisms in each seawater were counted by the number of colonies grown on the marine agar plate medium (cfu). More than tens of thousands of microbes were observed in the seawater, showing high biodegradability (Fig. 3).

The comparison of biodegradation activity between surface seawater and seawater of 5 m depth in Osaka is shown in Fig. 4. The biodegradation after 4 weeks was lower in the seawater at 5 m depth. This was consistent with the trend of the bacteria counts in seawater. The influence of the test temperature during the laboratory test was significant, and there is an optimum temperature. In general, biodegradation progresses most rapidly at around 30 °C. This suggests that the biodegradation proceeded faster in summer than in winter when compared to the same location in the actual environment. No significant difference was observed between the biodegradation activity of seawater at high and low tides, and the data did not show any detailed difference in the range of fluctuation during the day. In the comparison of the biodegradation activity of seawater during rainy and sunny days, the salinity of the seawater was also different, but there was no significant difference in the biodegradation activity. These facts indicate that the mass of seawater is very large and the quality of seawater is stable against daily changes in environmental conditions. However, the biodegradation activity tended to be higher immediately after typhoons and on days with high waves, and this is considered to be due to the effect of the roll-up of the seabed soil.

5 Polyamide 4 and Its Biodegradation

Biodegradable plastics are mainly biosynthetic and synthetic aliphatic polyesters, but they have a low melting point and limited strength because of "aliphatic". There are means to overcome these disadvantages by introducing aromatic groups or amide groups, which allow π-π or hydrogen bonding interactions between polymer chains. For example, copolymers of ε-caprolactone and ε-caprolactam (copolyesteramide) [9, 10], copolymers of amino acids and α-hydroxy acids (polydepsipeptide) [11],

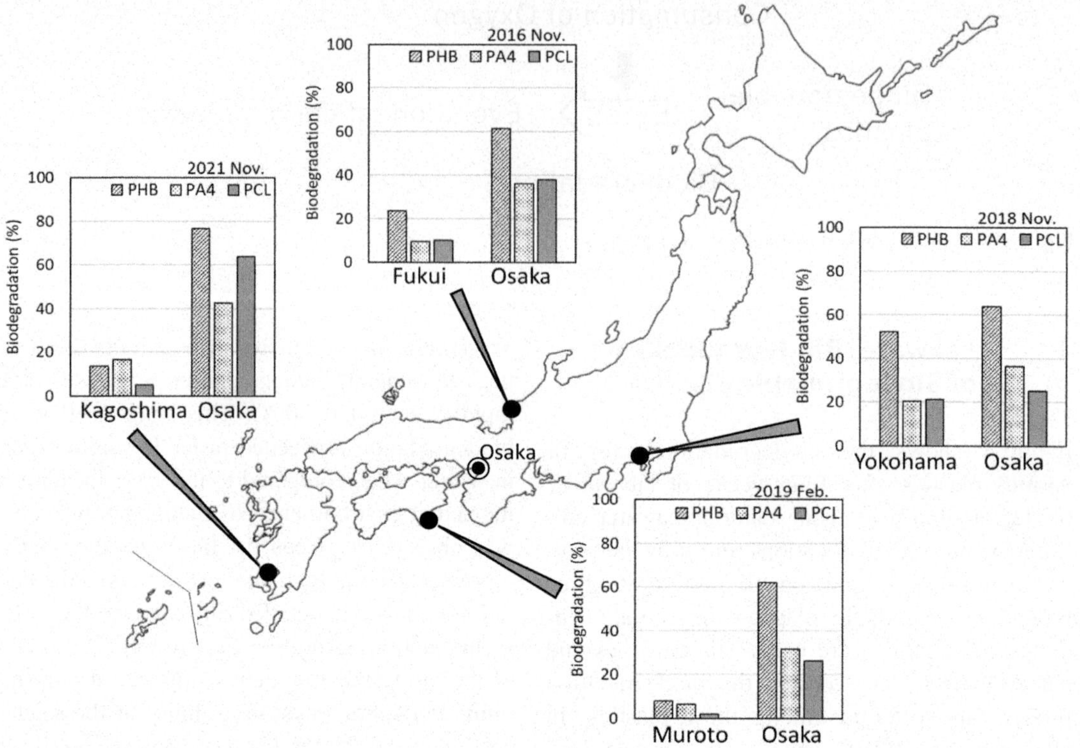

Fig. 3 Comparative biodegradation results using seawaters sampled at several places in Japan. Biodegradation tests of P3HB, PA4, and PCL powders were carried out for 4 weeks

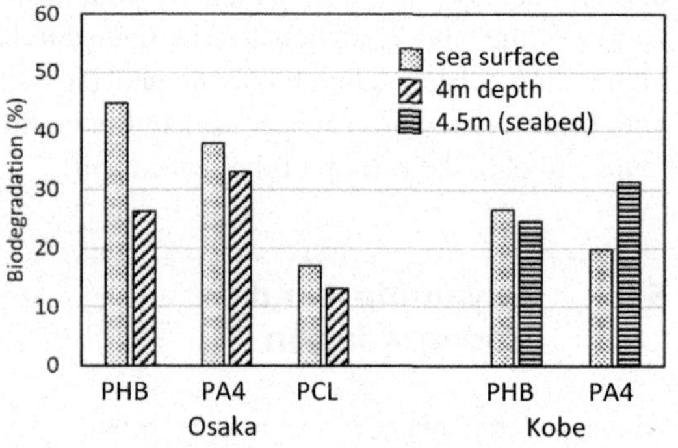

Fig. 4 Comparison of biodegradation activity of seawater sampled from sea surface and in the sea. Biodegradation results show the values after 4 weeks

and polyamino acids have been introduced [12]. Regarding the biodegradation of existing polyamides, the oxidative biodegradation of polyamides, PET and polyolefins by white rot fungi was reported [13]. However, since aromatic and amide groups have the effect of inhibiting biodegradability, there is a limit to providing both strength and biodegradability, and it is difficult to provide engineering plastic level strength and thermal properties maintaining biodegradability in a typical natural environment.

Polyamide 4 (PA4, nylon 4) has been attracting attention as a biodegradable plastic despite its structure without ester bonds in the main chain. In the course of investigating the relationship between the amide component ratio and biodegradability of a series of copolyesteramides, the authors found that the biodegradability of copolymers incorporating the polyamide 4 structure did not decrease even if the amide content was increased and that even polyamide 4, which is 100% amide, was biodegradable, which was the first finding from activated sludge biodegradation test.

PA4 is a polymer of 4-aminobutyric acid, which is two methylenes shorter than PA6 in the repeating unit. PA4 shows good biodegradability

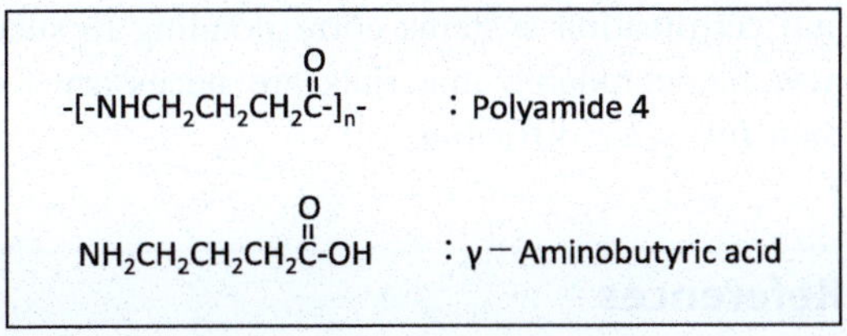

Fig. 5 Polyamide 4 and its monomer [16]

while PA6 and other nylons do not biodegrade at all. The biodegradability of PA4 is thought to be due to the fact that the polymer's building block, 4-aminobutyric acid (γ-aminobutyric acid, GABA), is an amino acid that occurs widely in nature and in living organisms, and enzymes that recognize this derivative occur widely in nature (Fig. 5) [14].

Polyamide 4 is a polyGABA, that is poly-amino acid. It is synthesized by ring-opening polymerization of a five-membered ring lactam (γ-butyrolactam) rather than by dehydration condensation of GABA [15]. 5-Membered lactam is a monomer with a stable structure and does not polymerize at high temperatures, it proceeds at temperatures below 50 °C. This monomer is synthesized from fossil resources through a multi-step process, but it can be easily obtained from biomass in a two-step process using glutamic acid as the raw material through GABA (Fig. 6) [14, 16].

The characteristic of PA4 is based on its short repeat unit length. Due to the relatively high amide group content in the repeating unit, PA4 is more hydrophilic than PA6 and has a higher water retention rate. The melting point and Vicat softening point are higher than those of PA6 (Table 3), and the heat resistance and tensile strength are greater than typical biodegradable polyesters. On the other hand, it has properties such as low dimensional stability due to water absorption, high processing temperature, and low elongation at break. PA4 is not a direct substitute for PA6, but is expected to be widely used as a biodegradable nylon with enhanced strength and heat resistance compared to conventional biodegradable polyester.

PA4 is biodegraded in the presence of various inoculum sources. For example, in the presence of activated sludge, PA4 powder was 60–80% biodegraded in about 5 weeks. With the progress of biodegradation, the concentration of NO_3 ion in the biodegradation bottle increased, and nitrogen in the polymer was also mineralized. In addition to PA4-degrading bacteria, activated sludge also contains many microorganisms that assimilate decomposition intermediates. Therefore, water-soluble intermediate degradation products do not accumulate in the test system [17]. PA4 biodegrades well in the natural environment, and a 100 μm film is completely

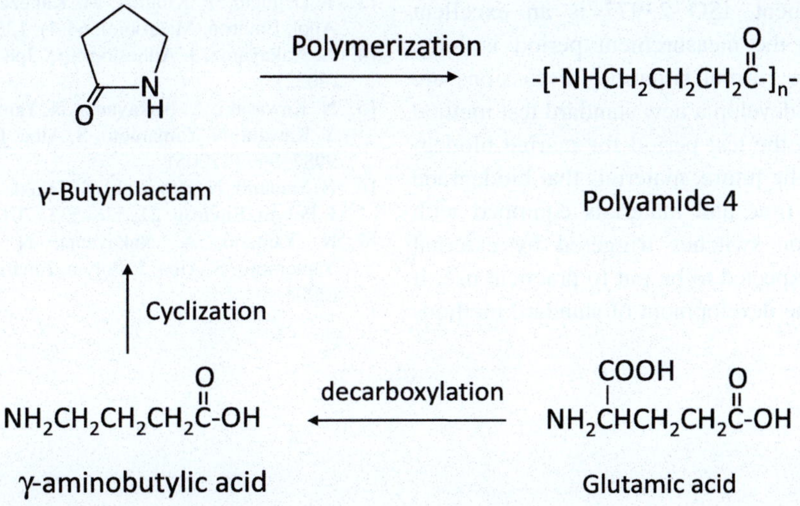

Fig. 6 Synthetic path of Polyamide 4 [16]

Table 3 Heat resistance of PA4 [16]

	PA4	PA6
Melting temp (°C)	265	215–225
Vicat softening temp. (°C)	236	213
Defelction temp. under flexural load (°C) 0.45 MPa	185	175

disappeared in about 60 days in the soil of the field in the summer season. Biodegradation of PA4 in seawater has already been introduced in the previous chapter (Figs. 3 and 4), and although the biodegradation rate is slower than biosynthetic resins such as P3HB, PA4 has excellent seawater biodegradability almost equal to that of PCL.

6 Conclusion

As the development of plastics showing good marine biodegradability is being promoted, it is urgent to improve the biodegradability evaluation method.

Methods related to marine biodegradation include ISO 18830, ISO 19679 and ISO 23977 laboratory test methods. The former uses seawater and sediment, but there is a problem that organic components brought in from the seabed soil become noisy and reduce the accuracy of measurement. ISO 23977 is an excellent method, but the measurement period is long. Based on the current situation, discussions are underway to develop a new standard test method that shortens the test period for marine biodegradation. In the future, materials that biodegrade slowly over time and materials equipped with biodegradation switches triggered by external stimuli are expected to be put to practical use. It seems that the development of standard methods and certification systems corresponding to such new biodegradable materials are necessary for their full-scale diffusion.

References

1. Biodegradable plastics and marine litter: Misconceptions, concerns and impacts on marine environments, reported by UNEP (2015)
2. N. Yamano, N. Kawasaki, M. Oshima, A. Nakayama, Polym. Deg. Sta. **108**, 116–122 (2014)
3. A. Nakayama, N. Kawasaki, I. Arvanitoyannis, J. Iyoda, N. Yamamoto, Polymer **36**(6), 1295–1301 (1995)
4. Y. Maeda, H. Mori, T. Maeda, O. Itoh, K. Yamaguchi, A. Nakayama, N. Kawasaki, N. Yamamoto, S. Aiba, J. Appl. Polym. Sci. **84**(10), 1838–1847 (2002)
5. N. Kawasaki, A. Nakayama, N. Yamano, S. Takeda, Y. Kawata, N. Yamamoto, S. Aiba, Polymer **46**, 9987–9993 (2005)
6. H. Sashiwa, R. Fukuda, T. Okura, S. Sato, A. Nakayama, Mar. Drugs **16**(1), 34–45 (2018)
7. A. Nakayama, N. Yamano, N. Kawasaki, Polym. Degrad. Stab. **166**, 290–299 (2019)
8. N. Yamano, N. Kawasaki, S. Ida, A. Nakayama, Polym. Degrad. Stab. **166**, 230–236 (2019)
9. Y. Tokiwa, T. Suzuki, T. Ando, J. Appl. Polym. Sci. **24**(7), 1701–1711 (1979)
10. Y. Tokiwa, T. Suzuki, J. Appl. Polym. Sci. **26**(2), 441–448 (1981)
11. Y. Ohya, M. Toyohara, M. Sasakawa, H. Arimura, T. Ouchi, Macromol. Biosci. **5**(4), 273–276 (2005)
12. A. Maruyama, N. Adachi, T. Takatsuki, M. Torii, K. Sanui, N. Ogata, Macromolecules **23**, 2748–2752 (1990)
13. T. Deguchi, Y. Kitaoka, M. Kakezawa, T. Nishida, Appl. Environ. Macrobiol. **64**(4), 1366 (1998)
14. A. Nakayama, J. Adhesion Soc. Jpn **57**(8), 338–344 (2022)
15. N. Kawasaki, A. Nakayama, N. Yamano, S. Takeda, Y. Kawata, N. Yamamoto, S. Aiba, Polymer **46**(23), 9987–9993 (2005)
16. N. Yamano, N. Kawasaki, S. Takeda, A. Nakayamal, J. Polym. Environ. **21**, 528–533 (2013)
17. N. Yamano, A. Nakayama, N. Kawasaki, N. Yamamoto, S. Aiba, J. Polym. Environ. **16**, 141–146 (2008)

Biodegradation Control of Ocean-Degradable Plastics by Photo-Switching

Akihiko Masui

1 Introduction

Plastics have been widely used as a material with characteristics such as light weight and excellent processability. However, as the use of fuels such as coal and petroleum is restricted in response to global warming issues, it has become necessary to move away from petroleum-based plastics. In addition, marine pollution caused by plastic waste has become an issue with the global increase in production since the 1990s.

It is thought that much of the plastic waste that is discharged into the environment eventually ends up in the ocean, and it is estimated that a total of 150 million tons of plastic waste has already been discharged into the world's oceans. It is estimated that at least 8 million tons of new plastic waste flow into the oceans every year [1]. The current situation is that we cannot keep up with the recovery of marine litter. This large amount of plastic waste is already having a significant impact on the marine ecosystem. For example, approximately 700 species of fish, seabirds, marine mammals, and sea turtles have been killed or injured by marine debris, 92% of which are due to the effects of plastic waste, such as the accidental ingestion of plastic bags [2].

According to the World Economic Forum, it is estimated that 82% of the marine plastic litter flowing into the ocean at present comes from Asian countries [3]. They predict that by 2050, the amount of plastic produced will quadruple, and the amount of marine plastic waste will exceed the amount of fish in the ocean [4]. Plastic litter is also causing huge economic losses. For example, it is estimated that the annual loss from plastic litter in the Asia Pacific region is $620 million per year in tourism and $360 million per year in fisheries [5].

Many types of plastic waste are degraded by physical effects such as waves and ultraviolet rays. Then, they become low molecular weight, mechanically brittle, disintegrated, crushed, and fragmented into small plastic particles, which are scattered in the marine environment around the world [6]. Plastic particles smaller than 5 mm are difficult to recover and are recognized as one of the environmental pollutants called microplastics. Reports of marine pollution by microplastics began in the 1970s [7]. Then, in response to the full-scale increase in plastic production, the number of reports related to microplastics has increased since the 2000s [3].

It is possible that microplastics may contain harmful substances because various chemical substances are added to plastics during

A. Masui (✉)
Osaka Research Institute of Industrial Science and Technology (ORIST), Izumi Center, 7-1 Ayumino-2, Izumi 594-1157, Osaka, Japan
e-mail: aki@orist.jp

T. Kaneko (ed.), *Photo-switched Biodegradation of Bioplastics in Marine Environments*,
https://doi.org/10.1007/978-981-99-4354-8_8

manufacturing to ensure practicality and moldability, or because new chemical substances are absorbed when they drift [8]. Microplastics in the world's oceans have been taken up by marine ecosystems, and at the same time, it has been pointed out that they may be contained in drinking water and salt [9]. Although the details of the effects of microplastics on the human body are still unknown, we cannot be optimistic about the consequences of introducing a substance that does not exist in nature into the body.

2 Biodegradable Plastics and Biodegradation Control

Biodegradable plastics have been researched and put to practical use for more than 30 years, because they do not become bulky waste after use, and are expected to reduce the volume of landfill space, as well as to be an environmentally low-impact material that will biodegrade and disappear when scattered in the environment [8, 10–14]. However, in addition to cost and market size, biodegradable plastics are subject to microbial action even during use, and biodegradation progresses moment by moment, resulting in a decrease in molecular weight and physical properties, making them unreliable as materials and an obstacle to their widespread use.

On the other hand, the degradability of the material is considered to be useful for reducing marine plastic litter. Although it would be ideal to replace all bottles, bags, and food containers that cause drifting litter with biodegradable plastics as a measure against marine plastic litter, it is not realistic. It is only necessary to reduce the amount of waste, collect it, and otherwise dispose of it appropriately, and to regulate and manage it so that it is not discharged into the ocean. On the other hand, in the case of fishing gear, which is a part of drifting litter, biodegradability is required because of the possibility of unintentional discharge into the ocean due to bad weather, natural disasters, operational accidents, and troubles, and the difficulty of recovery. In order to replace these fishing gears with biodegradable plastics, it is required that they do not biodegrade and maintain their physical properties for a certain period of time while being used in the ocean, and at the same time, it is desired that they degrade quickly after being discharged into the ocean. In other words, the switching function is required to maintain mechanical and other physical properties during use and to degrade quickly after use.

The physical properties can be enhanced by compositing various organic and inorganic nanofillers into the biodegradable plastic [15, 16]. Biodegradability can also be inhibited by mixing clays and metals [16–23]. However, these are all technologies to suppress biodegradation, and the degree of suppression cannot be controlled, and it does not degrade quickly after use.

3 Development of Photo-Switching Ocean-Degradable Plastics Utilizing the Antibacterial Function of Photocatalyst

Biodegradable plastics are needed in the field of fishing gears. In order to promote their widespread use, it is important to have a switching function, such as sufficient durability during use and biodegradation only after use. In order to provide this function, a trigger for biodegradation is necessary such as external factors including moisture, temperature, oxygen, pH, etc. However, it is difficult to make the trigger work after use. "Light" is expected to be a trigger for biodegradation because it is an environmental factor that easily reflects changes in the state of use, such as being exposed to indoor light or sunlight during use, and not being exposed to light in the soil or under the sea after use.

Photocatalysts, such as titanium dioxide (TiO_2), have been used in many fields because they exhibit high oxidation activity under ultraviolet and visible light [24, 25]. There have already been many research reports on composite with biodegradable plastics, but most of the purpose is to accelerate degradation by synergistic effect of biodegradation and

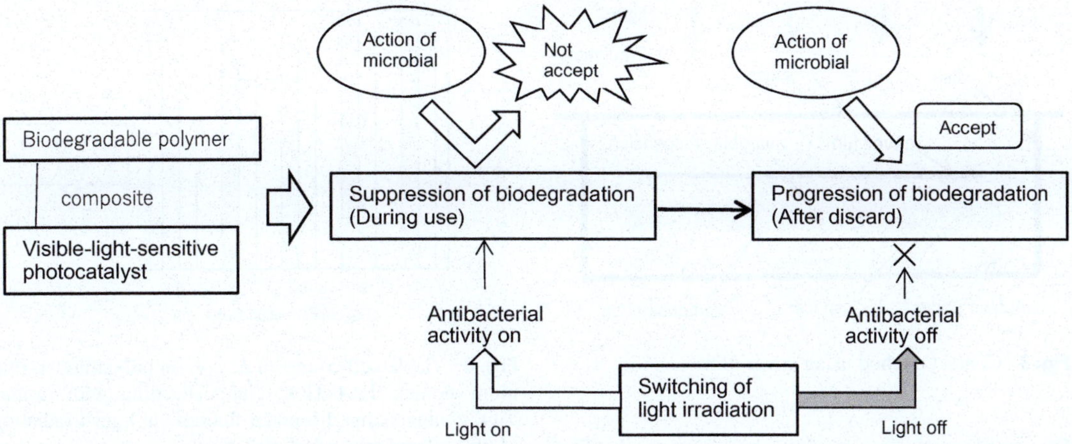

Fig. 1 Biodegradation control model in biodegradable polymer–visible-light-sensitive photocatalyst composite by on/off light irradiation

photodegradation by ultraviolet light [26–28]. On the other hand, photocatalyst also has antimicrobial activity, and if we utilize the antimicrobial activity by weak photoexcitation of modified photocatalyst, which has absorption band in the visible region, it is possible to suppress biodegradation during use (under light irradiation) and progress biodegradation after discard (in the dark) by using photo-switching function (Fig. 1).

Therefore, we composited visible-light-sensitive photocatalyst on biodegradable plastic and evaluated the antimicrobial activity (suppression of biodegradation) of the resulting composite film under light irradiation.

4 Evaluation of Antibacterial Activity of Photocatalytic Composite Film Under Light Irradiation

Polyamide 4 (PA4) was used as the biodegradable plastic. PA4 is a biobased material that can be easily synthesized from biomass via glutamate (Fig. 2) [29–37]. Nylons such as nylon 6 or nylon 6,6 are generally stable and difficult to degrade in natural environments. In contrast, PA4 is a type of nylon with distinct physical properties and yet it is a new material that can be easily biodegraded in the natural environment, such as activated sludge, soil, and seawater [33, 38–41]. Therefore, PA4 is one of the ideal candidates for practical biodegradable plastics.

The composite films were prepared by solvent cast method from 2,2,2-trifluoroethanol solution homogenized by ultrasonic treatment using PA4 and TiO_2 (copper compound including type) as visible-light-sensitive photocatalyst. The antibacterial activity of PA4–TiO_2 composite films was evaluated by the contact method using cover film (Fig. 3) as described in the International Organization for Standardization (ISO) 17094 [42]. The test bacteria were brought into contact with composite film (25×25 mm, 40 to 60 μm thickness) for a certain period of time under light irradiation (white fluorescent light: 1,000 lx to 4,000 lx) or in the dark (0.5 lx), and the increase or decrease in the viable bacteria cell counts after light irradiation were measured. The antibacterial efficacy (R) was evaluated by Eq. (1).

$$R = \log[B/C] \quad (1)$$

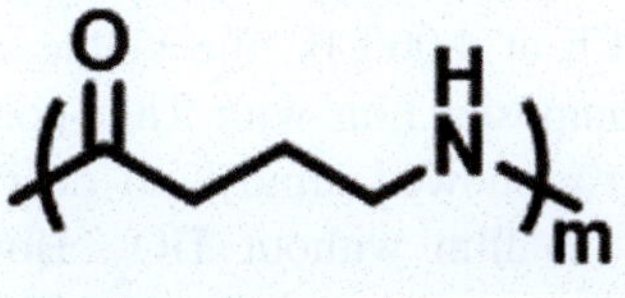

Fig. 2 Chemical structure of Polyamide 4 (PA4)

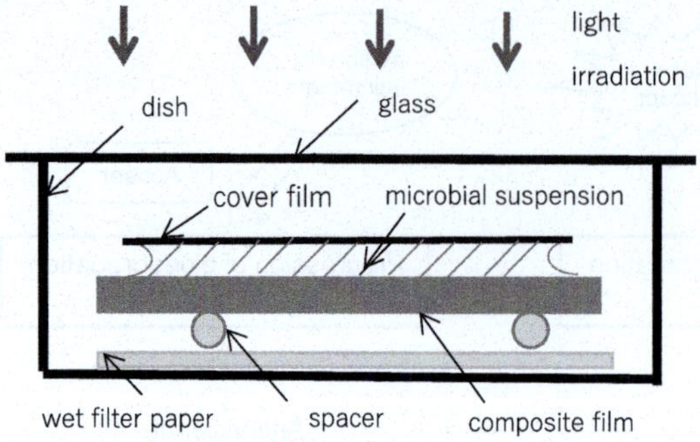

Fig. 3 Contact method using cover film

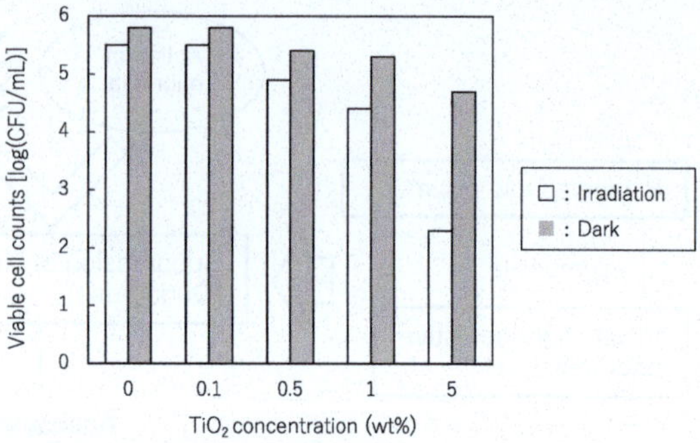

Fig. 4 Viable cell counts of *E. coli* on polyamide 4–titanium dioxide (PA4–TiO_2) composite films with various TiO_2 contents after 1 h under fluorescent light irradiation (4,000 lx) and in the dark (0.5 lx)

B = viable cell counts of the bacteria on the film without TiO_2.

C = viable cell counts of the bacteria on the film with TiO_2.

Escherichia coli NBRC 3972 (*E. coli*), *Staphylococcus aureus* ATCC 6538P (*S. aureus*), and PA4-degrading bacteria *Pseudomonas* sp. ND-11 [38] isolated from activated sludge were used as test microorganism to evaluate the antibacterial activity.

Scanning electron microscopic observation of the prepared composite films showed that the distribution of TiO_2 was different between the front (top) and back (bottom) surfaces of the films. More TiO_2 was distributed on the back surface. It is considered because TiO_2 sank downward in the polymer solution during casting. Therefore, since the evaluation of antibacterial activity by the contact method using cover film is based on the reaction between the composite film and the microorganisms on the contact surface, and whether the front or the back surface of the film is contacted may affect the results, all experiments were conducted with the back surface of the film in contact with the test bacteria to evaluate the antimicrobial activity [43].

The antibacterial activity of the composite film was investigated using *E. coli* as the test microorganism. After fluorescent light irradiation for 1 h at 4,000 lx, the viable cell counts on the composite film with TiO_2 concentration of 0.1 wt% showed almost no decrease compared to the film without TiO_2. However, the viable cell counts decreased as the concentration of TiO_2 increased, and the viable cell counts at TiO_2 concentration of 5 wt% decreased more than three orders of magnitude (R = 3.2). On the other hand, when the film was in contact with *E. coli* for 1 h in the dark, there was almost no decrease in the viable cell counts at TiO_2 concentration of 1 wt%, but a decrease of about one order of magnitude was observed on the film with TiO_2 concentration of 5 wt% due to the presence of copper compounds in the photocatalyst (Fig. 4; Table 1). Thus, the composite film showed a little antibacterial activity in the dark, but the antibacterial activity was dramatically enhanced by white fluorescent light irradiation.

In addition, we evaluated the antibacterial activity of the composite film with TiO_2 concentration of 0.5 wt% by changing the illumination intensity and irradiation time. The cell counts did not change compared to control after fluorescent light irradiation for 2 h at 2,000 lx. As the increase of the illumination and irradiation time, the viable cell counts gradually decreased. After 6 h of irradiation at 3,000 lx, the viable cell counts decreased by more than two orders of magnitude (R = 2.1) (Fig. 5; Table 2).

Table 1 Effect of TiO_2 content on the antibacterial efficacy of PA4–TiO_2 composite films against *E. coli* after irradiation for 1 h at 4,000 lx

TiO_2 content (wt%)	0.1	0.5	1	5
Antibacterial efficacy (R)	0.0	0.6	1.1	3.2

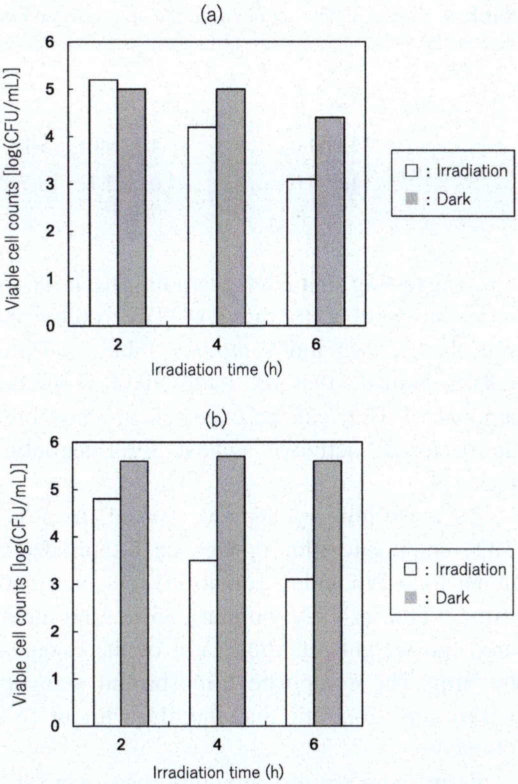

Fig. 5 Viable cell counts of *E. coli* on PA4–TiO_2 composite films with 0.5 wt% of TiO_2 after various times under irradiation (2,000 lx (**a**), 3,000 lx (**b**)) and in the dark

Table 2 Effect of illumination and irradiation time on the antibacterial efficacy of PA4–TiO_2 composite films against *E. coli*

Irradiation time (h)		2	4	6
Antibacterial efficacy (R)	2,000 lx	0.0	0.2	1.1
	3,000 lx	0.8	1.8	2.1

Thus, we were able to impart biodegradation control function in which high antibacterial activity is expressed by fluorescent light irradiation (suppression of biodegradation) and vanishingly low antibacterial activity is expressed in the dark (progression of biodegradation), to PA4 by compositing with TiO_2. Moreover, the antibacterial activity could be controlled by changing the TiO_2 concentration, illumination, and irradiation time.

The antibacterial activity of the composite films against *S. aureus* was evaluated by changing the TiO_2 concentration and irradiation time. After fluorescent light irradiation for 1 h at 4,000 lx, the viable cell counts showed no differences in any concentrations of TiO_2 in contrast to the case of *E. coli* (data not shown). When the irradiation time was increased to 4 h, a decrease in the viable cell counts was observed as the TiO_2 concentration increased, and the viable cell counts decreased by more than three orders of magnitude in the composite film with TiO_2 concentration of 5 wt% compared to the film without TiO_2 (R = 3.2) (Fig. 6; Table 3). *S. aureus* was more antibacterial resistant to the PA4–TiO_2 composite film than *E. coli* because the results of the 4-h irradiation of *S. aureus* were similar to those of the 1-h irradiation of *E. coli*. The biodegradation control function of the PA4–TiO_2 composite film could be applied to *S. aureus*, although the antibacterial activity was different compared to that of *E. coli*.

Furthermore, the same conditions as in the case of *E. coli* (4,000 lx, 1 h light irradiation) were used for the evaluation using ND-11 strain of PA4-degrading bacteria, and the viable cell counts became below the detection limit (10 CFU/ml) in the composite film with TiO_2 concentration of 5 wt% (data not shown). Therefore, we evaluated the antibacterial activity by decreasing the illumination intensity. When irradiated at 3,000 lx for 1 h, the viable cell

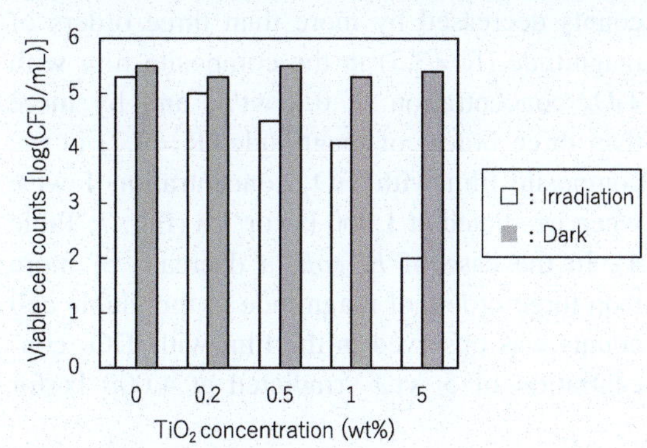

Fig. 6 Viable cell counts of *S. aureus* on PA4–TiO_2 composite films with various TiO_2 contents after 4 h under irradiation (4,000 lx) and in the dark

Table 3 Effect of TiO_2 content on the antibacterial efficacy of PA4–TiO_2 composite films against *S. aureus* after irradiation for 4 h at 4,000 lx

TiO_2 content (wt%)	0.2	0.5	1	5
Antibacterial efficacy (R)	0.2	0.7	1.1	3.2

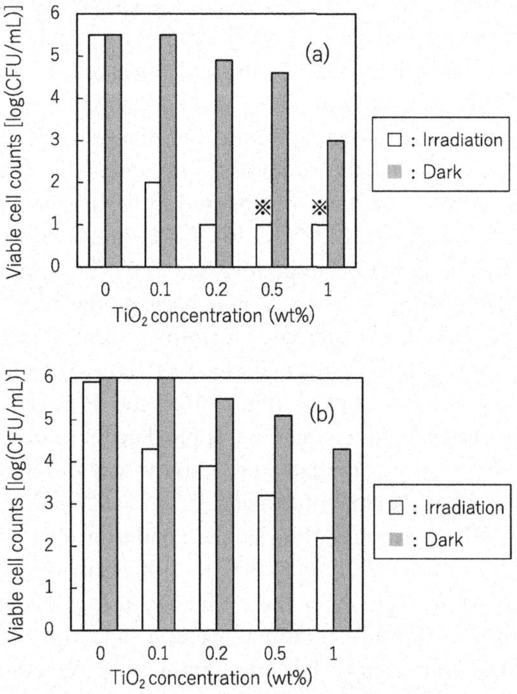

Fig. 7 Viable cell counts of *Pseudomonas* sp. ND-11 on PA4–TiO_2 composite films with various TiO_2 contents after 1 h under irradiation (3,000 lx (**a**), 1,000 lx (**b**)) and in the dark. ※: below detection limit (10 CFU)

counts decreased by more than three orders of magnitude (R = 3.3) in the composite film with TiO_2 concentration of 0.1 wt%, and by more than three orders of magnitude (R = 3.7) in the composite film with TiO_2 concentration 1 wt% when irradiated at 1,000 lx for 1 h (Fig. 7; Table 4). In the case of *E. coli*, a decrease of more than three orders of magnitude in the viable cell counts was observed in the film with TiO_2 concentration of 5 wt% irradiated at 4,000 lx for

Table 4 Effect of TiO_2 content on the antibacterial efficacy of PA4–TiO_2 composite films against *Pseudomonas* sp. ND-11

TiO_2 content (wt%)		0.1	0.2	0.5	1
Antibacterial efficacy (R)	3,000 lx	3.3	4.4	>4.5	>4.5
	1,000 lx	1.6	2.0	2.7	3.7

1 h, suggesting that PA4-degrading bacteria are more susceptible to the PA4–TiO_2 composite film than *E. coli* and *S. aureus* (Table 5). These results indicate that the addition of a smaller amount of TiO_2 can be expected to show high antibacterial activity against PA4-degrading bacteria.

As a preliminary test, we sinked the PA4–TiO_2 composite film in the sea (1.5 m depth, under light irradiation condition) for one month using a type of TiO_2 without copper, and measured the weight retention ratio of the composite film. The composite film showed a higher weight retention ratio than the film without TiO_2 (Table 6).

From these results, although there were differences in antibacterial activity depending on the type of microorganism, the biodegradability of photocatalytic composite film could be suppressed by the expression of antibacterial activity under light irradiation, and the biodegradability could be controlled by changing the amount of photocatalyst loading.

5 Future Subjects and Prospects

In this experiment, there was a slight decrease in the viable cell counts even in the dark when the concentration of TiO_2 was high because by using TiO_2 containing copper-based compounds. Therefore, it is necessary to develop a new photocatalyst that has sharp photo-switching

Table 5 The antibacterial efficacy of PA4–TiO_2 composite films against *E. coli*, *S. aureus*, and ND-11

TiO_2 content (wt%)			0.1	0.2	0.5	1	5
Antibacterial efficacy (R)	*E. coli*	4,000 lx, 1 h	0.0		0.6	1.1	3.2
	S. aureus	4,000 lx, 4 h		0.2	0.7	1.1	3.2
	ND-11	1,000 lx, 1 h	1.6	2.0	2.7	3.7	

Table 6 Weight retention ratio of PA4–TiO_2 composite films after sinking in the sea for one month

		TiO_2 (R type)		TiO_2 (A type)	
TiO_2 content (wt%)	0	1	5	1	5
Weight retention ratio (%)	76.5	81.4	83.7	76.9	80.4

ability and does not degrade the base material, biodegradable plastic itself. White fluorescent light specified by ISO was used as the light source In this experiment. However, considering practical use outdoors (under sunlight) such as in oceans and the difference of wavelength spectrum of the light source between white fluorescent light and sunlight (Fig. 8), it is necessary to use a light source that simulated sunlight such as xenon lamp as the light source in the future. In addition, biodegradable plastics include biosynthetic, chemically synthesized, and natural product-based materials. Each of them reacts with different degradable enzymes and different microorganism groups. Therefore, since the antibacterial activity of photocatalyst also varies depending on the biodegradable plastics, it is important to study the optimization of the amount of photocatalyst added and composited method.

Not only biodegradable plastics with a photo-switching function can be applied to fishing gears, but also they show antibacterial activity when exposed to intermittent weak light such as sunlight on land. For example, it is expected that the switching function is applied to mulch film in the field, because it will not decompose when it is spread out, and biodegradation will progress only after it is plowed into the soil. The photo-switching function is expected to be applied to a wide range of fields in agriculture, forestry, and fisheries. In the future, the systematization of knowledge on antibacterial control of target biodegradable plastic groups and microbial groups will lead to the development of various photo-switching biodegradable plastic products. Moreover, the composites can play a role of surface coater for bionylon products introduced in Chap. 3, in order to control photo-induced biodegradability of them.

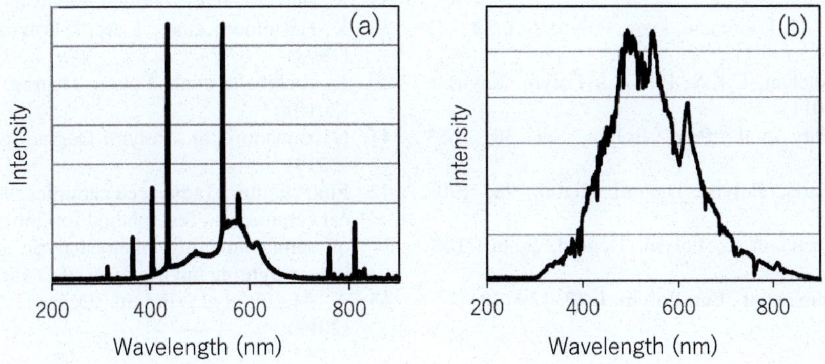

Fig. 8 Spectrum of fluorescent light (**a**) and sun light (**b**)

Acknowledgements This work was partly supported by JSPS Grants-in-Aid for Scientific Research 15K05593, NEDO Feasibility Study Program "Novel Marine Biodegradable Materials Based on Polyamide", and Grant-in-Aid from moon-shot Project (JPNP18016) "Development of photo-switching ocean-degradable plastics with edibility" of NEDO, Japan.

References

1. Stemming the Tide: Land-Based Strategies for a Plastic-Free Ocean. McKinsey & Company Report (2015)
2. S.C. Gall, R.C. Thompson, Mar. Pollut. Bull. **92**, 170 (2015)
3. J.R. Jambeck et al., Science **347**, 768 (2015)
4. The new plastics economy: rethinking the future of plastics. World Economic Forum Report (2016)
5. APEC Marine Resources Conservation Working Group, Economy Report—China (2009)
6. M. Eriksen et al., PLoS ONE **9**, e111913 (2014)
7. E.J. Carpenter et al., Science **177**, 85 (1972)
8. J P. Kershaw et al., Sources fate and effects of microplastics in the marine environment: part 2 of a global assessment. GESAMP Report (2015)
9. D. Yang et al., Environ. Sci. Technol. **49**, 13622 (2015)
10. H. Sashiwa, et al. Mar. Drugs **16** (2018). https://doi.org/10.3390/md16010034
11. D. Briassoulis, J. Polym. Environ. **12**, 65 (2004)
12. V. Siracusa et al., Trends Food Sci. Technol. **19**, 634 (2008)
13. L.S. Nair, C.T. Laurencin, Prog. Polym. Sci. **32**, 762 (2007)
14. G.E. Luckachan, C.K.S. Pillai, J. Polym. Environ. **19**, 637 (2011)
15. M.M. Reddy et al., Prog. Polym. Sci. **38**, 1653 (2013)
16. D.N. Bikiaris, Polym. Degrad. Stabil. **98**, 1908 (2013)
17. M. Nerantzaki et al., Polym. Degrad. Stabil. **108**, 257 (2014)
18. K. Fukushima et al., Eur. Polym. J. **47**, 139 (2011)
19. L.N. Luduena et al., J. Appl. Polym. Sci. **128**, 2648 (2013)
20. J.W. Rhim et al., Prog. Polym. Sci. **38**, 1629 (2013)
21. D. Mondal et al., J. Appl. Polym. Sci. **131**, 40079 (2014)
22. B. Tomsic et al., Polym. Degrad. Stabil. **96**, 1286 (2011)
23. D. Klemencic et al., Carb. Polym. **80**, 426 (2010)
24. K. Hashimoto et al., Jpn. J. Appl. Phys. **44**, 8269 (2005)
25. T. Ochiai, A. Fujishima, J. Photochem. Photobiol. **13**, 247 (2012)
26. M. Miyauchi et al., Environ. Sci. Technol. **42**, 4551 (2008)
27. S.P. Yew et al., Polym. Degrad. Stabil. **91**, 1800 (2006)
28. N. Sridewi et al., Polym. Degrad. Stabil. **91**, 2931 (2006)
29. V.A. Najjar, J. Fisher, Fed. Proc. **11**, 264 (1952)
30. R. Shukuya, G.W. Schwert, J. Biol. Chem. **235**, 1649 (1960)
31. A.D. Homola, E.E. Dekker, Biochemistry **6**, 2624 (1967)
32. N. Kawasaki et al., Polymer **46**, 9987 (2005)
33. N. Yamano et al., Polym. Degrad. Stabil. **108**, 116 (2014)
34. N. Kawasaki et al., J. Appl. Polym. Sci. **126**, 425 (2012)
35. N. Yamano et al., Polym. Degrad. Stabil. **137**, 281 (2017)
36. A. Nakayama et al., Polym. Degrad. Stabil. **98**, 1882 (2013)
37. N. Yamano et al., J. Polym. Environ. **21**, 528 (2013)
38. N. Yamano et al., J. Polym. Environ. **16**, 141 (2008)
39. K. Hashimoto et al., J. Appl. Polym. Sci. **54**, 1579 (1994)
40. K. Tachibana et al., Polym. Degrad. Stabil. **95**, 912 (2010)
41. N. Yamano et al., Polym. Degrad. Stabil. **166**, 230 (2019)
42. Fine ceramics (advanced ceramics, advanced technical ceramics)—Test method for antibacterial activity of semiconducting photocatalytic materials under indoor lighting environment. ISO 17094 (2014)
43. A. Masui et al., Polym. Degrad. Stabil. **167**, 44 (2019)

Structural Analysis of Nylon Hydrolase and Enzymatic Approach to Hydrolyze Polyamide Nylon

Dai-ichiro Kato, Naoki Shibata and Seiji Negoro

1 Introduction

Nylons are synthetic high molecular weight materials developed by DuPont in 1935 to replace natural silk. Two of the most famous nylons are nylon6, synthesized by ring-opening polymerization of ε-caprolactam, and nylon66, synthesized by condensed polymerization of adipic acid and hexamethylenediamine. Because of their toughness, flexibility, and heat- and chemical-resistance properties, these materials are used in daily life as fiber and plastic materials. The annual production of synthetic polyamides was more than 4 million tons in 2016, and only two nylons, nylon6, and nylon66, accounted for more than 90% of this total production.

D. Kato
Department of Science, Graduate School of Science and Engineering, Kagoshima University, Kagoshima, Japan
e-mail: kato@sci.kagoshima-u.ac.jp

N. Shibata
Department of Picobiology, Graduate School of Life Science, University of Hyogo, Himeji, Japan
e-mail: shibach@sci.u-hyogo.ac.jp

S. Negoro (✉)
Department of Applied Chemistry, Graduate School of Engineering, University of Hyogo, 2167 Shosha, Himeji 671-2280, Hyogo, Japan
e-mail: negoro@eng.u-hyogo.ac.jp

Considering that nylon materials are synthesized from fossil fuels, it is desirable to propose new methods of resource recycling circulation that do not place a burden on the environment as a countermeasure to global warming. White rot fungi are known as biocatalysts that exhibit nylon biodegradability, but their degradation strategy is methylene chain scission via radical reactions, and no method has been found to convert them back into reusable monomers [1]. Chemical recycling technology for nylon6 has been established, but for the conversion of nylon66 to constituting monomers (adipic acid and hexamethylenediamine), no effective methods have been reported so far [2].

Hydrolyzing polymeric nylons into monomers is not easy because of their unique properties. The toughness and flexibility of nylons are derived from an appropriate mixture of crystalline regions, where each polymer chain is regularly aligned and stabilized by intermolecular hydrogen bonds and amorphous regions, and each polymer chain is located in an irregular direction (Fig. 1). The crystallinity rate in solid nylon is approximately 20–50%, and physical properties such as melting point and strength depend on the degree of this crystallinity. Nylon is known to have two types of crystallization modes: α phase crystals in which the molecular chains are aligned inversely parallel to each other, and γ phase crystals in which the chains are aligned parallel to each other [3, 4]. These

T. Kaneko (ed.), *Photo-switched Biodegradation of Bioplastics in Marine Environments*,
https://doi.org/10.1007/978-981-99-4354-8_9

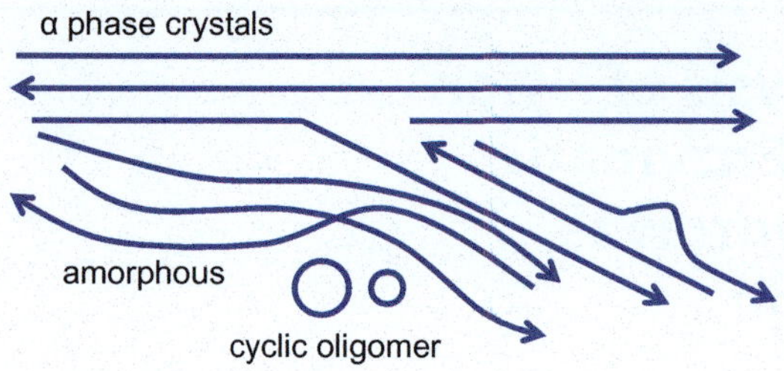

Fig. 1 Schematic diagram of 6-nylon structure and nylon oligomers. A mixture of α phase crystals (the molecular chains of adjacent polymers are aligned in an inverse parallel pattern) and amorphous regions, with a small amount of cyclic oligomers attached. Modified from Negoro et al. [19]

crystals make it difficult to expose the free amorphous amide bonds on the solid surface and therefore are less prone to hydrolysis.

Under these circumstances, we have isolated various microorganisms that degrade oligomers of 6-aminohexanoic acid (Ahx), which are produced during industrial nylon synthesis [5]. Ahx oligomers (nylon oligomers) are a general term for linear oligomers and/or cyclic oligomers after only a few polymerizations of ε-caprolactam, which is the raw material for the synthesis of nylon6, and so-called "low molecular weight nylons formed during polymerization" (Fig. 2a). We obtained approximately 10 species of bacteria that grow on Ahx oligomers as the sole carbon and nitrogen sources and conducted genetic and biochemical studies of the enzymes involved in Ahx oligomer degradation (Fig. 2b) [6–9]. Among them, *Arthrobacter* sp. KI72 was the first Ahx oligomer-degradable bacterium isolated in 1972. This strain has been deposited as NBRC 114184 in the Biological Resource Center, NITE (National Institute of Technology and Evaluation). The *Arthrobacter* bacterium possesses three enzymes showing different degradation modes for Ahx oligomers, i.e., 6-aminohexanoate-cyclic-dimer hydrolase (NylA, EC 3.5.2.12) [10], 6-aminohexanoate-dimer hydrolase (NylB, EC 3.5.1.46) [11–16], and 6-aminohexanoate-oligomer endohydrolase (NylC, EC 3.5.1. 117) [17, 18]. These enzymes are encoded on plasmid pOAD2, one of three plasmids harbored in strain KI72. NylA, NylB, and NylC are Ahx oligomer specific and have no detectable activity against the natural peptides tested. In addition, there is no immunological analogy or amino acid sequence homology with each other.

NylC exhibits endo-type hydrolytic activity toward linear or cyclic Ahx oligomers with a degree of polymerization of more than three (Fig. 2a). Recently, we found that the thermostability of wild-type NylC is drastically improved by small numbers of amino acid substitutions and that a thermostable NylC mutant shows hydrolytic activity on polymeric nylons even at ambient temperature [19]. In this paper, we describe the characteristics of NylC, the mechanism of its thermostabilization based on its three-dimensional structure, and the potential for industrial applications.

2 Characterization and Structure of NylC

The nylon (oligomer) hydrolase NylC has been found in *Arthrobacter* sp. KI72 (pOAD2), as well as in the alkaliphilic bacteria *Kocuria* sp. KY2, NBRC114282, and *Agromyces* sp. KY5R, NBRC114283. The corresponding genes are encoded on chromosomes in *Agromyces* and *Kocuria* species [8]. The NylC enzymes are referred to as $NylC_{p2}$, $NylC_A$, and $NylC_K$.

NylCs are classified into the N-terminal nucleophile hydrolase (N-tn) family based on the sequence homology of amino acids. These enzymes become mature proteins by autocleavage, which are initially expressed as inactive precursors, and the posttranslational autocleavage of the nascent polypeptide between Asn266 and Thr267 generates an active enzyme with an α subunit (27 kDa) and a β subunit (9 kDa) (Fig. 3a) [17, 18]. The newly generated N-terminal amino acid side chain in the β subunit becomes nucleophiles in catalytic reactions [20]. Thus, the nucleophilic N-terminal residue Thr267 in the β subunit is proposed to function as a catalytic residue for both autocleavage and substrate hydrolysis (Fig. 3b) [19]. Mutation experiments of NylC have confirmed that

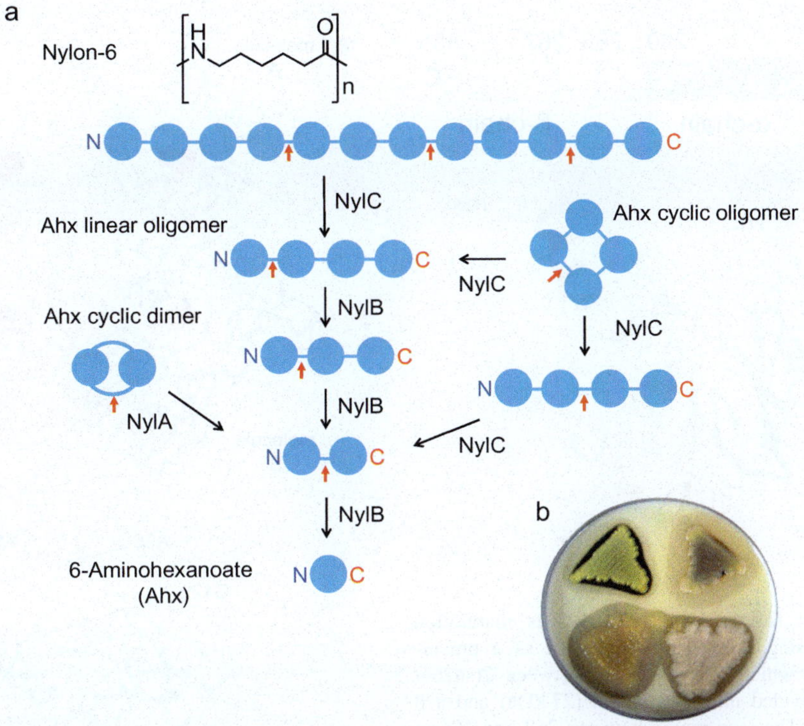

Fig. 2 Enzymatic hydrolysis of Ahx oligomers. **a** Three enzymes responsible for the degradation of nylon6, 6-aminohexanoate-cyclic, and linear oligomers to 6-aminohexanoate (Ahx) are schematically illustrated. NylA, 6-aminohexanoate-cyclic dimer hydrolase; NylB, 6-aminohexanoate dimer hydrolase; NylC, 6-aminohexanoate-oligomer endohydrolase (nylon hydrolase). The monomeric unit of Ahx is illustrated as a "light blue" circle. Terminal amino N and carboxyl C are shown as "light blue" and "red" letters, respectively. Red arrows indicate the hydrolytic sites catalyzed by each enzyme. **b** Growth of nylon oligomer-degrading microorganisms on an LB plate containing insoluble 6-aminohexanoate cyclic oligomers. Modified from Negoro et al. [5]

autocleavage is essential for the expression of enzyme activity.

In addition to Asn/Thr-type enzymes such as NylC, other autoprocessing recognition sequences are known to be Asn/Ser type (β-peptidylaminopeptidase, acylhomoserine lactone acylase), Asp/Thr type (glycosyl asparaginase), Gly/Ser type (D-aminopeptidase), Gly/Thr type (γ-glutamyltranspeptidase), Gly/Cys type (isopenicillin N-acyltransferase), and Ile/Cys type (N-acylsphingosine deacylase) [21–27].

X-ray crystallographic analysis revealed that four αβ heterodimers (molecules A-D) form a doughnut-shaped quaternary structure (Fig. 4a). Interestingly, the assembly of monomeric units (A-D) dynamically switches among the monomer, dimer (A/B or A/D), trimer, and tetramer (as well as higher oligomers), and the oligomeric states are affected by mutations, especially at the subunit interface (Figs. 4b, 5a). Wild-type $NylC_A$ and $NylC_K$ exist mainly as tetramers based on ultracentrifugation analysis [28]. In contrast, wild-type $NylC_{p2}$ was found to have a trimeric structure, in addition to monomeric and dimeric structures. The expected oligomeric states in aqueous solution, processability from the precursor to the active enzyme, thermostability, and enzyme activity are summarized in Table 1. We currently estimate that tetrameric structures are generated via the A/B dimer rather than the more unstable A/D dimer (Fig. 5a). In addition, small numbers of amino acid substitutions in NylC alter the successive fate of the nascent polypeptide (Fig. 5b). Four types of mutants have been obtained: active enzymes with altered thermostability or enzyme

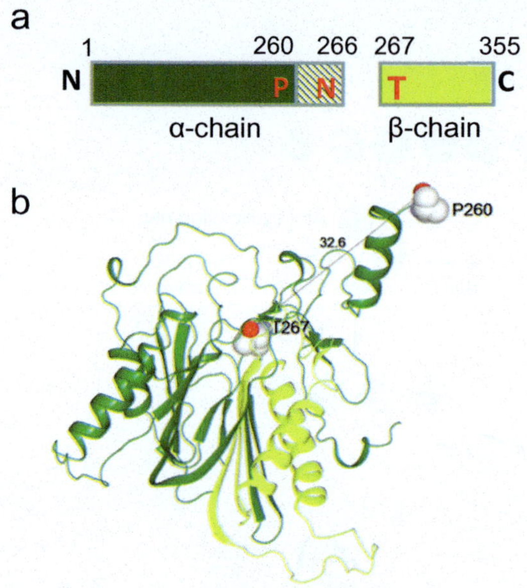

Fig. 3 Subunit structure of NylC and its maturation by autoprocessing. **a** NylC is expressed as a precursor (36 kDa), which is autoclaved between Asn266/Thr267 and divided into an α chain (27 kDa) and a β chain (9 kDa). The loop region between 260 and 266 of the α chain fluctuates greatly, and no electron density is observed in X-ray crystallographic analysis. **b** Subunit structure of NylC, in which the α (green) and β (yellow) chains form a heterodimer. Transported from Negoro et al. [19]

activity (type 1); soluble precursors (type 2); insoluble precursors (protein aggregation) (type 3); and fragments proteolyzed during the cultivation/purification process (type 4) [28].

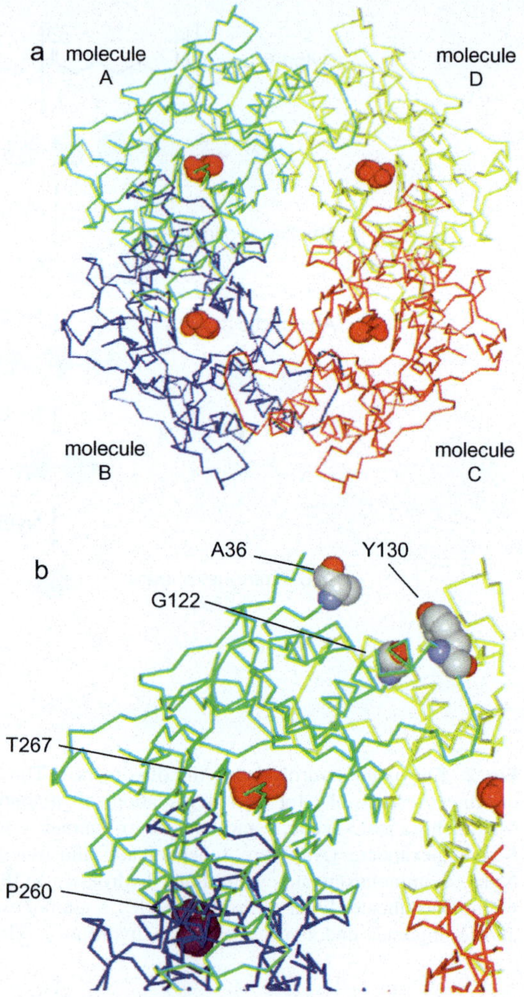

Fig. 4 Quaternary structure of GYAQ-NylC (PDB ID:5Y0M). **a** The quaternary structure is shown, with different colors highlighting the individual molecules A (*green*), B (*blue*), C (*red*), and D (*yellow*). The catalytic residue Thr-267 (the N-terminus of the β-subunit) is shown as a space-filling model (*red*). **b** An enlarged view of molecule A and its interfaces with adjacent molecules B and D. Three residues selected for mutagenesis, Ala-36, Gly-122, and Tyr-130, are shown as space-filling models. Because of the poor electron density distribution of the C-terminal region in the α-subunit, the adjacent Pro-260 is shown as a space-filling model (*magenta*)

3 Effect of Amino Acid Substitutions on Protein Thermostability

The amino acid sequence of $NylC_A$ differs from that of $NylC_{p2}$ in 5 places and that of $NylC_K$ differs in 10 more places in addition to the above 5 commons (Table 1) [8]. Thermal denaturation experiments using circular dichroism (CD) showed that the T_m values of $NylC_A$ and $NylC_K$ were 60 and 67 °C, respectively (Table 1; Fig. 6). These values are 8 and 15 °C higher than the T_m of $NylC_{p2}$ (52 °C). We have suggested that at least one of the five alterations (G111S, D122G, H130Y, L137A, V225M) in $NylC_A$ contributes to the increase in the thermostability of $NylC_{p2}$. Moreover, at least one of the 10 alterations (D36A, A41V, M50T, I60V, A62S, T230G, V231I, V257L, E263Q, G354A) in $NylC_K$ is estimated to contribute to a further increase in thermostability (Table 1).

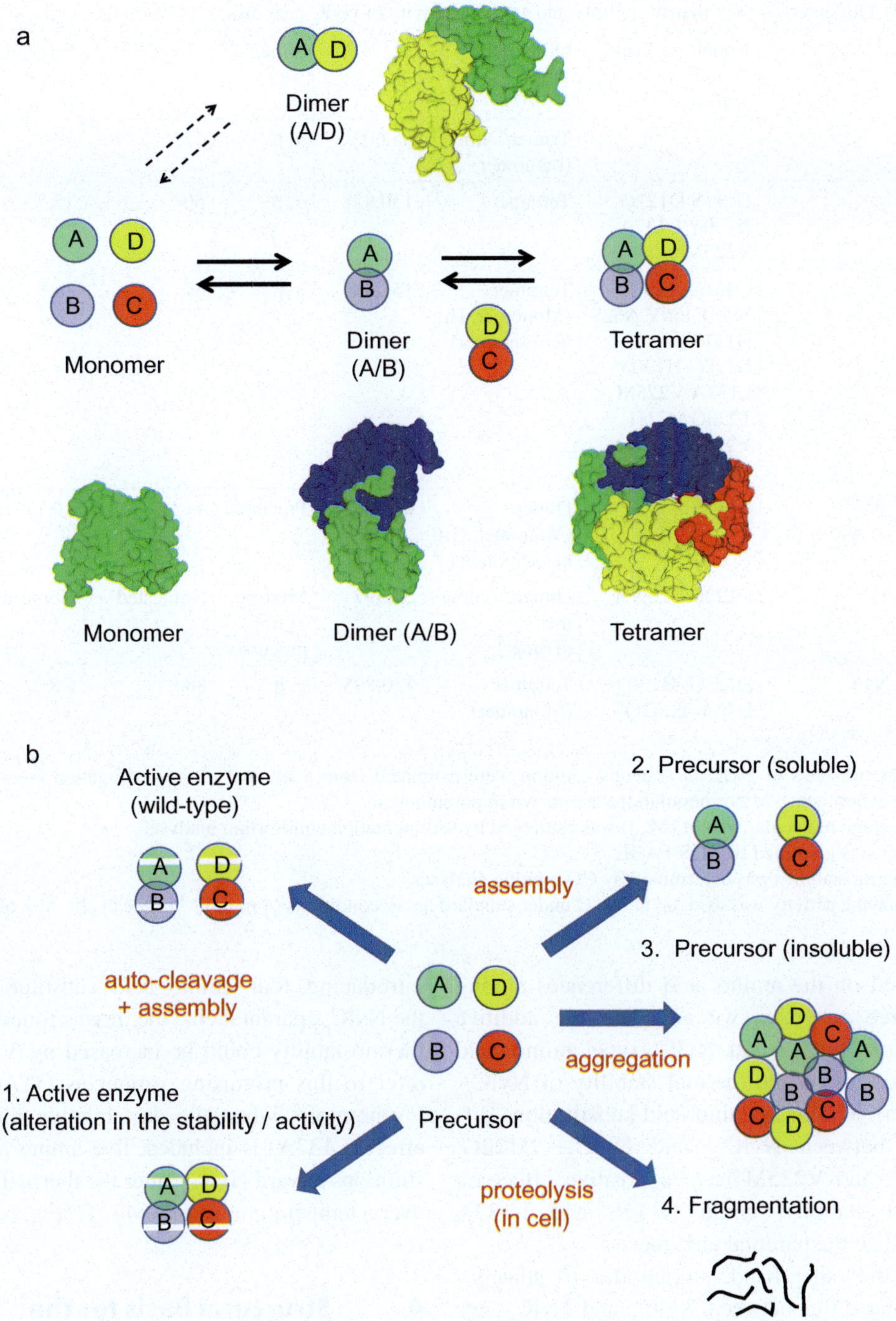

Fig. 5 Subunit assembly and mutational effects on expression of NylC. **a** Model of the subunit assembly of NylC. The oligomeric states are dynamically altered depending on the structural changes caused by mutations. Typical association/dissociation reactions are shown. Major assembly and dissociation reactions are assumed to be in equilibrium and are shown as solid lines, while the alternative equilibria shown as dashed lines may be possible. To simplify the designation of each monomer molecule in the quaternary structure, both the uncleaved precursor and the cleaved enzyme (αβ heterodimer) are expressed as molecules A-D. **b** Mutants that showed different assemblies were classified into four types. Type 1: obtained as active enzymes. Type 2: obtained as soluble precursors. Type 3: obtained as insoluble precursors (protein aggregation). Type 4: subjected to fragmentation during the cultivation/purification process. Transported from Negoro et al. [28]

Table 1 Oligomeric states, thermostability and enzyme activity of NylC enzymes

Enzyme	Amino acid substitutions in $NylC_{p2}$ sequence	Oligomeric state*1	M_{av} *2	Subunit*3	T_m (°C)*4	Enzyme activity*5 (U mg^{-1})
$NylC_{p2}$	–	Trimer/Dimer (monomer)	85,602	α, β	52	7.1
$NylC_A$	G111S D122G H130Y L137A V225M	Tetramer	130,928	α, β	60	15.7
$NylC_K$	D36A A41V M50T I60V A62S G111S D122G H130Y L137A V225M T230G V231I V257L E263Q G354A	Tetramer (Monomer, Higher oligomer)	151,384	α, β	67	14.9
$NylC_{p2}$-A^{137}	L137A	Dimer (Monomer, Higher oligomer)	60,258	Precursor	41	<0.1
$NylC_{p2}$-$G^{122}A^{137}$	D122G L137A	Dimer or monomer (Trimer)	67,397	Mixture of α, β, precursor	Not tested	Not tested
GYAQ-NylC ($NylC_{p2}$-$G^{122}Y^{130}A^{36}Q^{263}$)	D122G H130Y D36A E263Q	Tetramer (Monomer)	130,895	α, β	88	6.8

*1Oligomeric states of NylC in aqueous solution were estimated from analytical ultracentrifugation (sedimentation velocity experiment). Minor populations are shown in parentheses
*2The average molecular weight (M_{AV}) was estimated by sedimentation equilibrium analysis
*3Subunit was identified by SDS-PAGE
*4The thermostability was determined by CD melting analyses
*5The enzyme activity was assayed at 30 °C under standard assay conditions (4 mg ml^{-1} of the cyclic Ahx oligomer)

Based on the amino acid differences among the three enzymes, we analyzed the additive effect of $NylC_A$- and $NylC_K$-type amino acid substitutions on the thermal stability of $NylC_{p2}$ (Fig. 6a). Within 5 amino acid substitutions differing between $NylC_{p2}$ and $NylC_A$, D122G, H130Y, and V225M have a positive effect on thermal stability, while G111S and L137A destabilize the parental enzymes.

On the other hand, among the 10 substitutions that differ between $NylC_A$ and $NylC_K$, we focused on E263Q and D36A, located at the A/B and A/D subunit interfaces, respectively. When these two mutations were introduced into the D122G/H130Y double mutant (T_m=81 °C), the T_m value increased to 88 °C, yielding the highest thermal stability quadruple mutant (D36A/D122G/H130Y/E263Q). In other words, by introducing four amino acid substitutions into the $NylC_{p2}$ parental enzyme, it was found that the thermostability could be increased by 36 °C. We refer to this promising variant as GYAQ-NylC. Furthermore, when the destabilizing mutational effect (L137A) is included, five amino acid substitutions toward $NylC_{p2}$ alter the thermal stability over a temperature range of 47 °C (Fig. 6) [19].

4 Structural Basis for the Enhanced Thermostability by Mutations

The thermostability of the three NylCs was significantly changed by amino acid alternations occurring at the subunit interfaces. For example, D122G, H130Y, and D36A are located at the

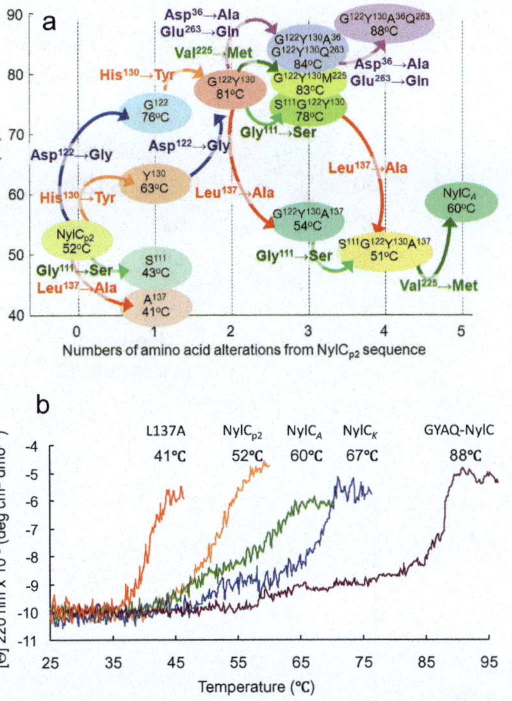

Fig. 6 Thermostability of NylC mutants. **a** The cumulative effects of amino acid mutations at the A/D subunit interface (D122G, H130Y, L137A, G111S, D36A) and A/B subunit interface (V225M, E263Q) on the melting temperatures of denaturation (T_m) are shown. **b** Thermal transition curves of the various NylC mutant enzymes. CD measurements were performed at 220 nm from 25 to 95 °C (1 °C min^{-1}). The results are expressed as the mean residue molar ellipticity [θ]. Protein concentrations of 0.1 mg ml^{-1} were used. *deg*, degree. Modified from Negoro et al. [19]

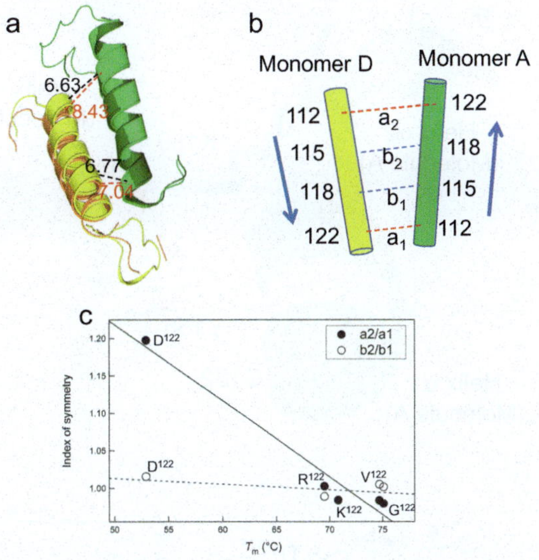

Fig. 7 Subunit interactions between monomers A and D. **a** The structure of helix α1 and its adjacent loop regions in monomers A and D is shown as a ribbon diagram. The structure of NylCp2-G122 (monomer A, *green*; monomer D, *yellow*, PDB ID: 5XYO) is superimposed on the structure of NylCp2 (monomer A, *dark green*; monomer D, *orange*, PDB ID:5XYG). **b** The relationship between the α1 helices in monomers A and D is illustrated. a1, distance between position 112 (monomer A) and position 122 (monomer D) at Cα; a2, distance between position 122 (monomer A) and position 112 (monomer D); b1, distance between position 115 (monomer A) and position 118 (monomer D) at Cα; b2, distance between position 118 (monomer A) and position 115 (monomer D). **c** The index of symmetry (a2/a1 and b2/b1) is plotted against the protein stability (melting temperature: T_m) estimated from CD analysis of NylCp2 (wild-type) and NylCp2-R122, K122, V122, and G122 mutants. Modified from Negoro et al. [28]

A/D interface, and E263Q is located at the A/B interface (Fig. 4b). Based on the three-dimensional structure of NylC, we estimate that the following structural effects cause changes in the protein stability and oligomeric states.

To evaluate the local asymmetry in the oligomeric structures, we measured the relationship between the α1 helix in monomers A and D. The T_m value estimated from the CD exhibited an inverse correlation with the index of symmetry (a_2/a_1 ratio) in the wild-type (Asp122) and mutant (Arg122, Lys122, Val122, and Gly122) enzymes, in which a_1 is the distance between position 112 (monomer A) and position 122 (monomer D) at C_α and a_2 is the distance between position 122 (monomer A) and position 112 (monomer D) (Fig. 7) [28]. In particular, in NylC$_{p2}$, the partial asymmetry at the A/D interface inhibits tetramer formation but generates an irregular trimeric structure in aqueous solution (Table 1). In addition, a clearer correlation was found between thermostability and the nearest distance between the side chain atoms on the encountered two helices [28]. In wild-type NylC$_{p2}$, Lys159-NH_3^+ (in molecule A) was 2.87 Å away from Glu115-$O_{\varepsilon 1}^-$ (in molecule D). In contrast, the distance was reduced to 2.65 Å in the thermostable Gly122 enzyme (Fig. 8a). This distance decreases with an almost linear correlation with increasing thermostability in the mutant enzymes (Fig. 8c). We

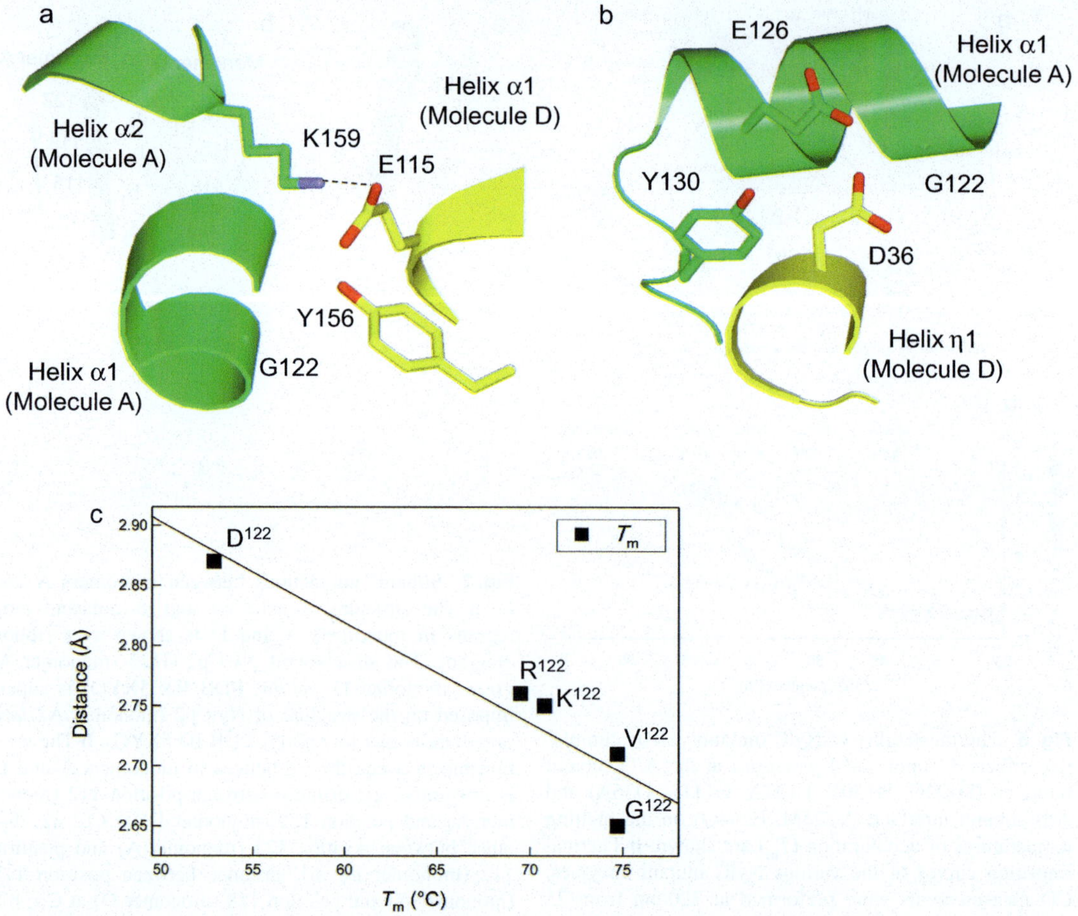

Fig. 8 Possible interactions at A/D subunit interface of $NylC_{p2}$-G122-Y130 (PDB ID: 5Y0L). **a, b** The possible interactions at the A/D subunit interface around Gly-122 (located on helix a1) (*A*), Asp-36 (located on 3_{10} helix h1)/Tyr-130 (located on loop region L3) (*B*), are shown. Molecule A and molecule D in the quaternary structure are colored *green* and *yellow*, respectively. C. The distance from Lys159-$NH3^+$ (located on helix α2 in monomer A) to Glu115-$O\varepsilon_1^-$ (located on helix α1 in monomer D), which is indicated as *dotted lines* in section A, is plotted against the T_m. Modified from Negoro et al. [28]

currently estimate that the electrostatic effect between the two residues will enhance subunit binding around Gly122. However, in $NylC_{p2}$, the amino acid residue at position 122 is replaced with Asp, and the close proximity of the acidic residue Asp122 (close to Glu115) reduces this stabilization effect.

Y130 (in molecule A) is located at a loop region (R127-G133) between helix α1 and 3_{10} helix η3. Y130-Oη forms a hydrogen bond network involving D36-N of the adjacent subunit (Fig. 8b). Mutating Y130 to histidine (H) may destroy hydrogen bonding network and destabilize the subunit interaction. Therefore, an H130Y substitution in $NylC_{p2}$ will contribute to an increase in thermostability by stabilizing the subunit interaction.

D36 (in molecule D) is located at 3_{10} helix η1, which is close to E126 (in molecule A) (4.40 Å), whereas D36 is replaced with A36 in $NylC_K$ (Fig. 8b). Therefore, the D36A mutation will reduce electrostatic repulsion with E126. Actually, D36A substitution (in G122-Y130 and G122-Y130-Q263) improves the thermostability.

E263 is located at the terminal region of the α-subunit (V261-N266), which had a poor electron density distribution in the X-ray diffraction study of wild-type and all NylC mutant enzymes tested. In GYAQ-NylC, the acidic amino acid residue E263 is replaced with neutral glutamine (Q), suggesting that altering the electrostatic environment induced by the E263Q mutation assists in improving subunit interactions (Fig. 4b).

5 Enzymatic Hydrolysis of Polyamide Nylon6

The nylon polymer chains are stabilized by forming hydrogen bonds with adjacent polymer chains aligned. Some chains, however, do not align and constitute amorphous regions. Nylon hydrolase is proposed to attack the amorphous region and hydrolyze amide bonds. However, if some amide bonds in the polymer aggregate are broken, the surrounding amide bonds still form hydrogen bonds with other molecules; therefore, the oligomer should not be released from the solid phase. In other words, the polymer is expected to be resistant to the hydrolytic reaction until the nick number (threshold for disintegration) is exceeded, at which point the degradation products are continuously released.

Among our constructed NylC mutants, a quadruple mutant, GYAQ-NylC, shows the highest thermal stability ($T_m = 88$ °C). Kinetic studies of GYAQ-NylC with the Ahx cyclic oligomer determined that the k_{cat} and K_m values were 2.8 ± 0.11 s^{-1} and 0.72 ± 0.06 mg ml^{-1}, respectively. When compared with the kinetic parameters of wild-type $NylC_{p2}$ ($k_{cat} = 6.5 \pm 0.29$ s^{-1}; $K_m = 3.7 \pm 0.27$ mg ml^{-1}), these mutations decreased the turnover of the product but improved the affinity of the enzyme for substrate [8]. As a result, the calculated k_{cat}/K_m of GYAQ-NylC (3.9 s^{-1} ml mg^{-1}) was twofold higher than that of the $NylC_{p2}$ WT enzyme (1.8 s^{-1} ml mg^{-1}). Based on the high thermostability and improved catalytic activity of GYAQ-NylC, we considered this mutant to be suitable for testing the hydrolysis of polyamide nylon6.

To detect the nylon degradation of powdered nylons, mechanically disintegrated nylon powder from nylon 6 pellets was subjected to the reaction at 30 °C. Gel permeation chromatography (GPC) is generally used to measure the molecular weight of polyamide nylon. However, the reaction for the solid polymer proceeds at the polymer surface. Therefore, it is likely that the molecular weights of the residual particles (inner regions) are not significantly affected, even after soluble reaction products are released from the solid samples. Considering this possibility, we attempted to apply gas cluster secondary ion mass spectrometry (GC-SIMS), which detects the alteration of mass sizes at the solid surface. In GC-SIMS, the sample is irradiated with low-energy cluster argon to excite it, which enables ionization under milder conditions. When the molecular weight distributions of the nylon sample before and after the enzymatic reaction with GYAQ-NylC were measured, the peak after the reaction shifted to a lower molecular weight range, confirming that GYAQ-NylC was clearly degrading nylon6 and nylon66 (Fig. 9) [19].

Oligomers released into the solution were quantified by detecting the amino group colorimetric method using trinitrobenzene sulfonic acid (TNBS method). However, the degradation rate of nylon6 and nylon66, which was the ratio of amino groups theoretically released to those generated when all amide bonds were hydrolyzed, was found to be limited to only 1.3% (Fig. 10). Moreover, although the amount of free amino groups increased to double the value by raising the reaction temperature from 30 °C to 60 °C, its value was still not satisfactory. This was thought to be because the sites at which the enzyme could attack were limited, retaining strong hydrogen bonds with the adjacent polymer chains and having few amorphous regions. Thus, for the next step, we planned to design the following two types of polymers to increase the enzymatic degradation.

The first is the heterogenous monomer insertions that disrupt the hydrogen bonds between polymer molecules. We created copolymers with monomer

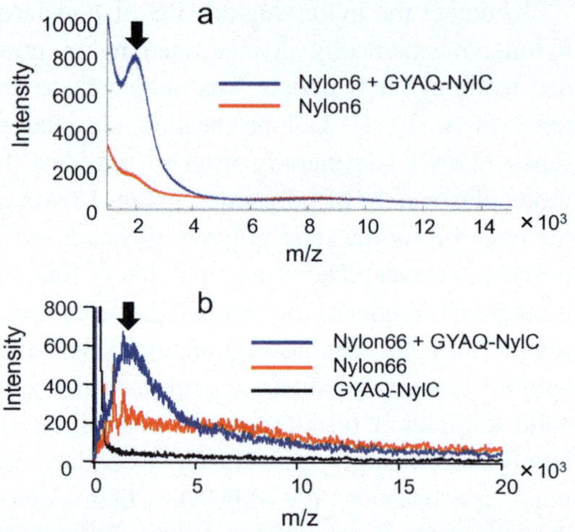

Fig. 9 Nylon degradation tests using argon cluster SIMS. Nylon6 or nylon66 powder (10 mg) was pretreated in triplicate in buffer A (180 μl) at 120 °C for 20 min. GYAQ-NylC (1 mg ml^{-1}, 20 μl) was then added and incubated at 60 °C for 2 h. The solid fraction was washed with distilled water, lyophilized, and dissolved in trifluoroethanol (0.2 ml), and a fraction (0.02 ml) was spotted onto a silicon plate (1 cm^2). The *m/z* range of 1,500–20,000 was analyzed. The experiment was performed in triplicate. The m/z of the arrow peak increases after the enzymatic reaction. **a** Nylon6; *blue line*, GYAQ-NylC-treated nylon6; *red line*, nylon6 alone. **b** Nylon66; *blue line*, GYAQ-NylC-treated nylon66; *red* line, nylon66 alone; *black* line, GYAQ-NylC alone. Modified from Negoro et al. [19]

insertions that disrupt the hydrogen bonds between each molecule. For instance, four carbon units were added to the nylon66 polymerization reaction in the range of 0.12–0.32 mol equivalent succinic acid. When GYAQ-NylC was applied, the decomposition rate increased as the percentage of succinic acid monomer increased [29].

The second is the construction of high-strength polymers with a low degree of polymerization. By lowering the degree of polymerization, the threshold for disintegration is expected to be lowered, and the decomposition speed is expected to increase. Therefore, we performed chemical degradation of nylon6 using formic acid and evaluated the degradation rate using GYAQ-NylC (Fig. 10). Actually, the degradation rate increased with the duration of the formic acid treatment, with the 50-h-treated sample (degree of polymerization: 30–40) showing a degradation rate of more than 40%. On the other hand, there was no significant change in the polymer melting point even when the degree of polymerization was reduced to approximately 30, which led us to expect that it would be possible to construct a polymer with an improved degradation rate. However, GYAQ-NylC cannot degrade Ahx dimers. Therefore, to achieve complete conversion to monomers, another enzyme, NylB, which cleaves linear oligomers into exo-type oligomers, was added to the GYAQ-NylC enzyme. As expected, the degradation rate was dramatically improved, and the rate of degradation after 50 h of formic acid treatment reached 84% (Fig. 10).

6 Quantification of Degradation Velocity Using Thin Nylon Film

Measuring the increase or decrease of substrate or product is commonly utilized to evaluate the reaction speed. However, the size of the nylon particles obtained by the conventional freezing disruption method varies, and their surface area is not constant, making it difficult to determine the active substrate concentration and thus the reaction rate. Therefore, we established a measurement system for the hydrolytic activity using a nylon film in which the contact area between the enzyme and the substrate is always constant [29].

In the experiment, the reflected light intensity was quantified from an image of a thin nylon

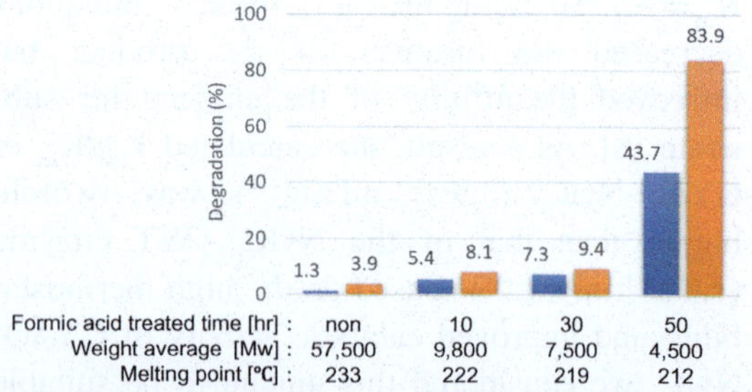

Fig. 10 Degradation of formic acid-treated nylon6. GYAQ-NylC alone (blue) and GYAQ-NylC + NylB combination system (yellow)

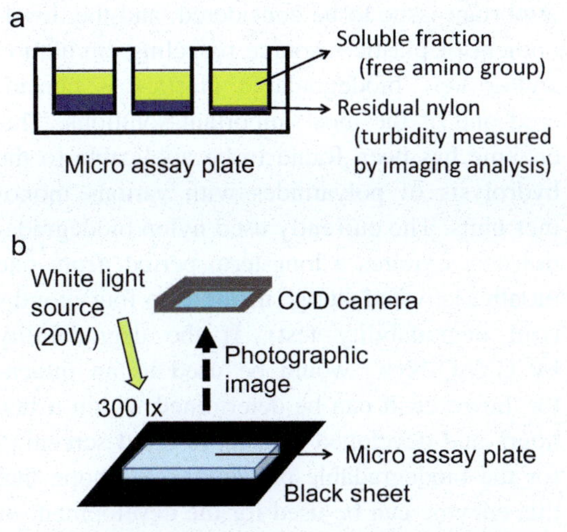

Fig. 11 Assay system using nylon layer immobilized on microassay plate. **a** Model of the nylon layer in a microassay plate. The hydrolysis of nylon was detected by the release of amino groups (soluble fraction) and changes in the turbidity of the nylon layers (insoluble fraction). **b** Estimation of the thickness of the nylon layer by imaging analysis. Photographic images of the thin layer obtained by the CCD camera were analyzed by image analysis programs. Modified from Nagai et al. [29]

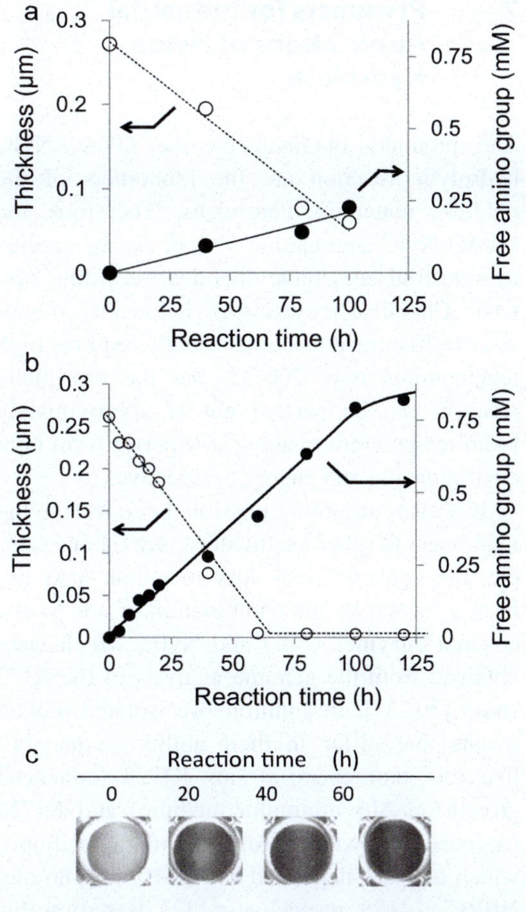

Fig. 12 Time course of degradation of thin-layered nylon-(66-co-64(0.32)). **a**, **b** Enzyme reactions were initiated by adding GYAQ-NylC (250 μl) to 5–11 wells of a microassay plate, where the synthesized nylon-(66-co-64(0.32) was attached as thin layers. After incubating the plates at 30 °C, all of the soluble fractions in each well were recovered at different incubation periods (4–150 h). Enzyme reactions were monitored by densitometric analysis of the thin layer (insoluble fraction) (*open blue circle*) and release of free amino groups (soluble fraction) (*red circle*). The following two enzyme concentrations were used: A 1.0 mg ml^{-1}; B 3.0 mg ml^{-1}. **c** Continuous change in the appearance of nylon-(66-co-64(0.32)) thin films treated by GYAQ-NylC degradation. Modified from Nagai et al. [29]

film taken by an imaging analyzer equipped with a CCD camera. As a result, we found that there was a correlation between the measured film thickness and the reflected light intensity (Fig. 11). By this property, the change in film thickness could be calculated by enzymatic reaction. We also estimated the layer number of the nylon film by inferring the distance between polymer chains based on information on the crystal structure and crystallinity of nylon. For example, a nylon6 thin film with a thickness of 260 nm could be estimated to have a layer of 600 nylon polymers.

Figure 12 shows the results of the determination of free amino groups by the TNBS method and the change in film thickness of nylon6 and succinic acid copolymer of nylon66 (nylon66-co-64(0.32)) thin films. As expected, the amount of free amino groups increased at a constant rate, and the film thickness decreased linearly. At an enzyme concentration of 1 mg/ml, the film degradation rate was calculated to be 2.2 nm/h (Fig. 12a). An increase in the amount of free amino groups also was observed in correlation with the decrease in the thickness of the nylon film. Thus, this measurement system has a high potential to be used as a high-throughput screening for the research development of biodegradable polyamides and nylon hydrolases with enhanced enzymatic activity.

7 Prospects for Industrial Applications of Nylon Hydrolase

The products obtained by the GYAQ-NylC hydrolytic reaction are the monomers of the polymer materials themselves. Therefore, the GYAQ-NylC degradation system can be applied to industrial enzymatic chemical recycling systems. Chemical recycling of nylon into monomers by chemical reactions usually requires high temperatures over 200 °C, but the enzymatic reaction can be carried out at approximately room temperature, making it superior from both environmental and energy perspectives.

It is also becoming possible to convert nylon monomers to other useful substances. For example, the conversion of Ahx to adipic acid has been achieved by the combination of newly discovered enzymes NylD and NylE, which were obtained from the genome analysis of the KI72 strain [30, 31]. In addition, we isolated mutant strains that differ in their ability to metabolize Ahx monomers: strains KI723 (enhanced growth on Ahx-minimum medium) and KI724 (decreased growth on Ahx-minimum medium), which also are deposited as NBRC 114186 and NBRC 114188, respectively [32]. Based on this information, we have identified the enzymes involved in monomer metabolism, and we are forecasting to establish a technology to convert nylon into organic acids for industrial utilization.

On the other hand, in the textile industry, enzymatic fiber processing has been used for a long time. Some cellulases are used as texture modifiers for cotton and rayon fibers, and protease is also used as a wool texture modifier [33]. On the other hand, there have been no practical examples of nylon fiber processing. This is because there have been no industrial enzymes that can practically hydrolyze nylon fibers. Because of the high thermostability of GYAQ-NylC, it is expected to be applied to the surface modification of nylon fibers through industrial enzymatic reactions.

As a countermeasure against global warming, the reduction of environmental impact is an important issue to be considered, and the development of plastic resource recycling circulation and/or new biodegradable plastics is considered one of the most important solutions. This enzyme has been found to be applicable to the hydrolysis of polyamides with various monomer units. The currently used nylon biodegradation test requires a long-term period, from one month (activated sludge method) to four months (soil degradability test). If the degradability by GYAQ-NylC would be used as an indicator, however, it can be determined within a few hours and developed as a more rapid screening for the biodegradable polyamide. We hope that this enzyme can be used for the development of novel biodegradable polyamides such as itaconic acid-derived bionylons introduced in Chap. 3 and their recycling in the future.

Funding This work was partly supported by the Japan Society for the Promotion of Science KAKENHI (19K05171) from the Ministry of Education, Culture, Sports, Science and Technology-Japan to S.N., and Moonshot Research and Development Program (JPNP18016) commissioned by the New Energy and Industrial Technology Development Organization (NEDO) to D.K.

References

1. M. Fujisawa, H. Hirai, T. Nishida, J. Polym. Environ. **9**, 103 (2001)
2. A. Kamimura, S. Yamamoto, Org. Lett. **9**, 2533 (2007)
3. S. Dasgupta, W.B. Hammond, W.A. Goddard, J. Am. Chem. Soc. **118**, 12291 (1996)
4. S. Saeid Arabnejad, S. Manzhos, AIP Adv. **5**, 107123 (2015)
5. S. Negoro, D. Kato, T. Ohki, K. Yasuhira, Y. Kawashima, K. Nagai, M. Takeo, N. Shibata, K. Kamiya, Y. Shigeta, Methods Enzymol. **648**, 357 (2021)
6. S. Negoro, Appl. Microbiol. Biotechnol. **54**, 461 (2000)
7. K. Kato, K. Ohtsuki, Y. Koda, T. Maekawa, T. Yomo, S. Negoro, I. Urabe, Microbiology **141**, 2585 (1995)
8. K. Yasuhira, Y. Tanaka, H. Shibata, Y. Kawashima, A. Ohara, D. Kato, M. Takeo, S. Negoro, Appl. Environ. Microbiol. **73**, 7099 (2007)

9. K. Yasuhira, Y. Uedo, M. Takeo, D. Kato, S. Negoro, J. Biosci. Bioeng. **104**, 521 (2007)
10. K. Yasuhira, N. Shibata, G. Mongami, Y. Uedo, Y. Atsumi, Y. Kawashima, A. Hibino, Y. Tanaka, Y.-H. Lee, D. Kato, M. Takeo, Y. Higuchi, S. Negoro, J. Biol. Chem. **285,** 1239 (2010)
11. S. Negoro, T. Ohki, N. Shibata, N. Mizuno, Y. Wakitani, J. Tsurukame, K. Matsumoto, I. Kawamoto, M. Takeo, Y. Higuchi, J. Biol. Chem. **280**, 39644 (2005)
12. T. Ohki, Y. Wakitani, M. Takeo, K. Yasuhira, N. Shibata, Y. Higuchi, S. Negoro, FEBS Lett. **580**, 5054 (2006)
13. S. Negoro, T. Ohki, N. Shibata, K. Sasa, H. Hayashi, H. Nakano, K. Yasuhira, D. Kato, M. Takeo, Y. Higuchi, J. Mol. Biol. **370**, 142 (2007)
14. T. Ohki, N. Shibata, Y. Higuchi, Y. Kawashima, M. Takeo, D. Kato, S. Negoro, Protein Sci. **18**, 1662 (2009)
15. Y. Kawashima, T. Ohki, N. Shibata, Y. Higuchi, Y. Wakitani, Y. Matsuura, Y. Nakata, M. Takeo, D. Kato, S. Negoro, FEBS J. **276**, 2547 (2009)
16. Y. Kawashima, K. Yasuhira, N. Shibata, Y. Matsuura, Y. Tanaka, M. Taniguchi, Y. Miyoshi, M. Takeo, D. Kato, Y. Higuchi, S.J. Negoro, Mol. Cat. B: Enzym. **64**, 81 (2010)
17. S. Kakudo, S. Negoro, I. Urabe, H. Okada, Appl. Environ. Microbiol. **59**, 3978 (1993)
18. S. Negoro, S. Kakudo, I. Urabe, H. Okada, J. Bacteriol. **174**, 7948 (1992)
19. S. Negoro, N. Shibata, Y. Tanaka, K. Yasuhira, H. Shibata, H. Hashimoto, Y.-H. Lee, S. Oshima, R. Santa, S. Oshima, K. Mochiji, Y. Goto, T. Ikegami, K. Nagai, D. Kato, M. Takeo, Y. Higuchi, J. Biol. Chem. **287**, 5079 (2012)
20. H. Suzuki, Kumagai. H. J. Biol. Chem. **277**, 43536 (2002)
21. L. Fanuel, C. Goffin, A. Cheggour, B. Devreese, G. Van Driessche, B. Joris, J. Van Beeumen, J.-M. Fre're, J. Biochem. J. **341**, 147 (1999)
22. B. Geueke, T. Heck, M. Limbach, V. Nesatyy, D. Seebach, H.P. Kohler, FEBS J. **273**, 5261 (2006)
23. J.M. Elkins, N.J. Kershaw, C.J. Schofield, Biochem. J. **385**, 565 (2005)
24. R. Sankaranarayanan, M.M. Cherney, C. Garen, G. Garen, C. Niu, M. Yuan, M.N. James, J. Mol. Biol. **397**, 979 (2010)
25. T. Okada, H. Suzuki, K. Wada, H. Kumagai, K. Fukuyama, Proc. Natl. Acad. Sci. USA **103**, 6471 (2006)
26. F. Marc, P. Weigel, C. Legrain, N. Glansdorff, V. Sakanyan, J. Biol. Chem. **276**, 25404 (2001)
27. K. Michalska, A. Hernandez-Santoyo, M. Jaskolski, J. Biol. Chem. **283**, 13388 (2008)
28. S. Negoro, N. Shibata, Y.-H. Lee, I. Takehara, R. Kinugasa, K. Nagai, Y. Tanaka, D. Kato, M. Takeo, Y. Goto, Y. Higuchi, Sci. Rep. **8**, 9725 (2018)
29. K. Nagai, K. Iida, K. Shimizu, R. Kinugasa, M. Izumi, D. Kato, M. Takeo, K. Mochiji, S. Negoro, Appl. Microbiol. Biotechnol. **98**, 8751 (2014)
30. I. Takehara, D. Kato, M. Takeo, N. Seiji, Genome Announc. **5**, e00217–e317 (2017)
31. I. Takehara, T. Fujii, Y. Tanimoto, D. Kato, M. Takeo, S. Negoro, Appl. Microbiol. Biotechnol. **102**, 801 (2018)
32. S. Negoro, H. Shinagawa, A. Nakata, S. Kinoshita, T. Hatozaki, H. Okada, J. Bacteriol. **143**, 238 (1980)
33. J. Hu, D. Tian, S. Renneckar, J.N. Saddler, Sci. Rep. **8**, 3195 (2018)

Ecotoxicity Assessment of Biodegradable Plastics in Marine Environments

Yoshifumi Horie and Hideo Okamura

Abbreviations

BPs: Biodegradable and bio-based plastics

MP	Microplastics
PLA	Polylactic acid
PBS	Polybutylene succinate
PBSA	Polybutylene succinate adipate
PBAT	Polybutylene adipate terephthalate
PHA	Polyhydroxy alkanoates
PHB	Polyhydroxy butyrate

CPs: Conventional petroleum-based plastics

HDPE	High-density polyethylene
LDPE	Low-density polyethylene
PS	Polystyrene
PP	Polypropylene
PET	Polyethylene terephthalate
PVC	Polyvinyl chloride
PUR	Polyurethane

Ecotoxicological assessment

ASTM	American Society for Testing and Materials
AF	Assessment factor
A/F	Antifouling
DMSO	Dimethyl sulfoxide
EC_{50}	50% Effective concentration
EQS	Environmental quality standard
ISO	International organization for standardization
LC_{50}	50% Lethal concentration
LOEC	Lowest observed effect concentration
NOEC	No observed effect concentration
OECD	Organization for economic cooperation and development
PEC	Predicted environmental concentration
PNEC	Predicted no effect concentration
SSD	Species sensitivity distribution
TG	Test guidelines

Y. Horie · H. Okamura (✉)
Research Center for Inland Seas, Kobe University,
Fukaeminami 5-1-1 Higashinada, Kobe 658-0022,
Japan
e-mail: okamurah@maritime.kobe-u.ac.jp

Y. Horie
e-mail: horie@people.kobe-u.ac.jp

T. Kaneko (ed.), *Photo-switched Biodegradation of Bioplastics in Marine Environments*,
https://doi.org/10.1007/978-981-99-4354-8_10

1 Risk Assessment of Chemicals

1.1 Basic Concept

Chemical substances used in land such as pesticides and fertilizers pass from fields through rivers, and administered to the human body such as pharmaceuticals and personal care products also pass from toilets through sewage treatment plants and rivers, and finally reach the sea. Plastics, on the other hand, are indispensable for our daily life and industrial activity, but are unintentionally discarded as waste and eventually reach the sea through rivers. Thus, all kinds of wastes with/without hazard chemicals enter the sea, and a wide variety of chemical substances are present in seawater at low concentrations.

In aquatic effects assessment, the low-risk concentration where no unacceptable adverse effects on the ecosystem are expected, i.e., PNEC is calculated, and it is compared with the concentrations that are present in the environment, either measured or calculated, i.e., PEC. When the PEC exceeds the PNEC, further assessment or risk management action needs to be considered [1]. Environmental risk assessment of chemicals consists of three procedures: exposure assessment, hazard assessment, and risk characterization. Exposure assessment is a procedure to obtain PEC, and hazard assessment is used to obtain PNEC. The ratio PEC/PNEC is used as a quantitative index for the risk assessment. For example, A/F biocides are used for fishing nets and submerged structures as well as for ship hulls to prevent biofouling on purpose. A risk assessment method that protects the marine environment from the potential negative impacts of biocidally active substances which are intentionally used for the A/F systems on ships is specified as a standardized methodology by ISO 13073-1 [2].

1.2 Ecotoxicity Testing to Determine PNEC

Standardized ecotoxicity tests to determine PNEC have been proposed by several international organizations. The OECD Guidelines for the Testing of Chemicals is a collection of about 150 of the most relevant internationally agreed testing methods used by government, industry, and independent laboratories to identify and characterize potential hazards of chemicals. In the guidelines, several organisms in different trophic levels such as decomposers, producers, and consumers are selected as recommended test species, ranging in bacteria, algae, crustaceans, mussels, fish, soil microorganisms, earthworms, crops, bees, and birds, to assess the ecotoxicity of chemicals on biotic systems, however, major aquatic organisms recommended are freshwater species. ASTM International and ISO also have proposed standardized ecotoxicity tests to assess polluted water and sediments. As toxic values EC50 and LC50 are obtained from acute test and NOEC and LOEC from chronic test.

1.2.1 Effects on Bacteria and Algae

ISO has proposed standardized ecotoxicity tests using bacteria, microalgae, and macroalgae to assess seawater quality. Inhibitory effect of water samples on the light emission of luminescent bacteria *Vibrio fischeri* after 30–60 min exposure [3–5] is toxicity endpoint. Inhibition of water samples on the growth of microalgae *Skeletonema* sp. and *Phaeodactylum tricornutum* after three days exposure [6] and the marine and brackish water macroalga *Ceramium tenuicorne* after seven days exposure [7] is toxicity endpoint.

1.2.2 Effects on Invertebrates

ISO and ASTM have proposed several ecotoxicity tests to assess seawater and sediment quality using various invertebrates with lethality, embryo-larval development, and abnormalities in life-cycle as toxicity endpoints. The following tests have been proposed, acute lethal toxicity tests to marine copepods [8], life-cycle toxicity tests with saltwater mysids [9], short-term chronic toxicity tests with echinoid embryos [10], acute toxicity tests to the marine rotifer *Brachionus plicatilis* [11], acute, chronic, and life-cycle aquatic toxicity tests with polychaetous annelids [12], static short-term chronic toxicity tests starting with embryos of four species of saltwater bivalve mollusks [13], toxicity of water with embryo-larval development of Japanese oyster *Crassostrea gigas* and mussel *Mytilus edulis* or *Mytilus galloprovincialis* [14], toxicity of sediment-associated contaminants with estuarine and marine invertebrates [15].

1.2.3 Effects on Fish

Several tests are currently available for examining the toxicity of chemicals in fish, specifically to determine the impacts of chemical exposure on growth, sexual development, and reproduction, as well as effects across multiple generations. Examples of such tests within the Organization for Economic Cooperation and Development Guidelines for the Testing of Chemicals (OECD TG) include TG 203: Fish, Acute Toxicity Test [16]; TG 212: Fish, Short-term Toxicity Test on Embryo and Sac-Fry Stages [17]; TG 210: Fish, Early-life Stage Toxicity Test [18]; TG 234: Fish Sexual Development Test [19]; TG 229: Fish Short Term Reproduction Assay [20]; and TG 240: Medaka Extended One Generation Reproduction Test [21]. The time needed to complete these tests ranges from 94 h for TG 203 to around 5 months for TG 240. The fish most commonly used for toxicity testing are the Japanese medaka (*Oryzias latipes*), zebrafish (*Danio rerio*), and fathead minnow (*Pimephales promelas*), which are useful as model animals because of their small genome size, small body size, and short generation time under small-scale breeding conditions.

In 2010, as part of efforts to reduce the use of animals in research, the European Union recommended the use of fish embryos as a more humane form of toxicity testing [22]. Then, in 2013, EU Directive 2010/63/EU on the protection of animals used for scientific purposes [22] was enacted, in which larvae immediately after hatching but before feeding behavior are not subject to the usual animal welfare legislation. In response to this directive, the Fish Embryo Acute Toxicity Test (TG 236) was developed for determining the lethal effect of chemicals in embryonic-stage zebrafish [23].

1.3 Example of Environmental Risk Assessment

Antifouling paints including A/F compounds are used to prevent biofouling on ship hulls and fishing nets. Cybutryne, same as Irgarol 1051, is one of seventeen antifouling compounds used in Japan (Japan Paint Manufactures Association, https://www.toryo.or.jp/jp/anzen/af/index.html, 2019). Cybutryne has been used as antifouling biocide worldwide for about 30 years as its residue was firstly reported in Mediterranean seawater in 1993, and its antifouling usage will be prohibited from 1 January 2023 according to IMO website (https://www.imo.org/en/MediaCentre/MeetingSummaries/Pages/MEPC-75th-session.aspx, 2022). Considerable data on ecotoxicological hazard and environmental persistence have been accumulated on this compound. Martins et al. [24] summarized all the ecotoxicity data on organic and organo-metallic antifouling co-biocides including cybutryne, to estimate each EQS using a workflow diagram as shown in Fig. 1. Ecotoxicological data set were collected from non-standard toxicity tests as well as standard tests, and the toxicity values in references were arranged for freshwater and marine/estuarine species. Deterministic PNEC was calculated by dividing the lowest toxicity values (EC50 and LC50 from acute test, NOEC and LOEC from chronic test) by assessment factor. Probabilistic PNEC was calculated by dividing the hazard concentration 5% from SSD analysis with toxicity values (EC50 and LC50 from acute test, NOEC and LOEC from chronic test) by assessment factor. Total 90 toxicity values from 57 marine species and 33 freshwater species such as bacteria, microalgae, macrophyte, zooplankton, and fish were used to calculate PNEC and EQS of cybutryne (Table 1). EQS_{SW} and EQS_{FW} of cybutryne were proposed as 1.4 ng/L and 0.22 ng/L, respectively, which were same as the lowest refined SSD obtained from chronic test. The data indicate that freshwater organisms are about 6 times more sensitive than marine species toward cybutryne. Residue concentrations in water

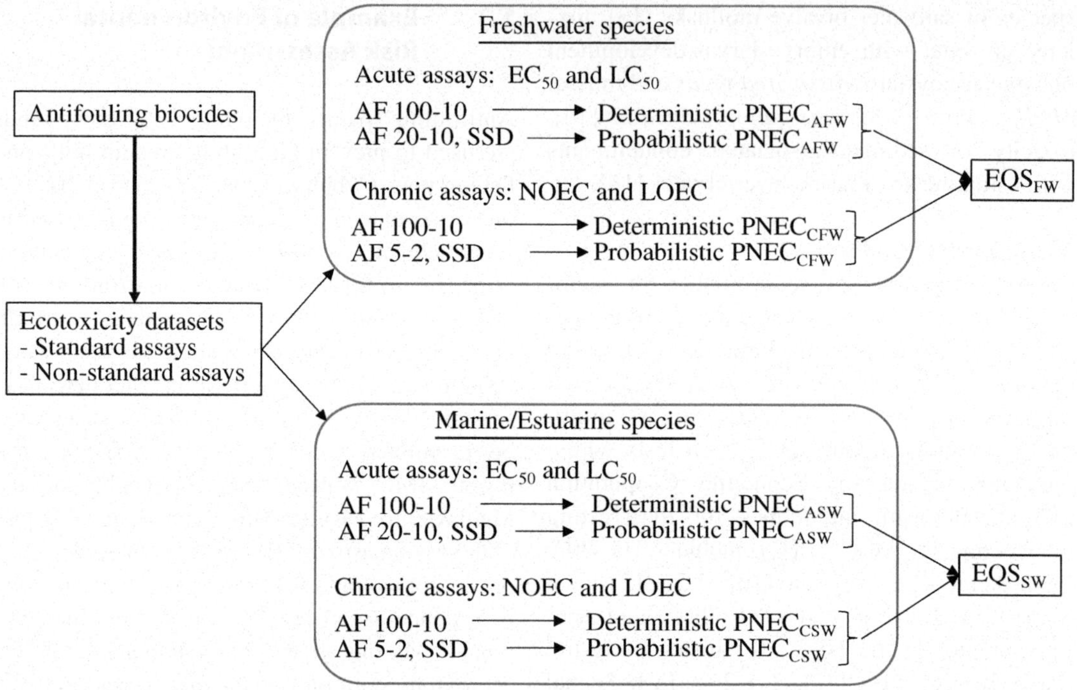

Fig. 1 Workflow diagram to determine EQS of antifouling biocide. Original figure from Martins et al. Biofouling 34, 34–52 (2018). AF: assessment factor, EQS: environmental quality standard, PNEC: predicted no observed effect concentration, EC: effective concentration, LC: lethal concentration, SSD: species sensitivity distribution, FW: freshwater, SW: seawater

were collected from 12 references [25–36] and a percentile curve composed of total 327 monitoring data is indicated in Fig. 2. All the residue concentrations of cybutryne exceeds proposed EQS_{FW} and 99% exceeded proposed EQS_{SW}. The data indicate that cybutryne has the potential to cause adverse effects on aquatic organisms. Thus, the methodology of environmental risk assessment of cybutryne could be applicable to BPs as well, where the PEC and PNEC of the stable degradation products from the BP are needed.

2 Ecotoxicity of BPs, CPs, and Related Compounds

All plastic materials can be further fragmented by wave and photochemical actions once they enter the aquatic environments to produce smaller pieces. CPs are reviewed on the hazard associated with additives like phthalates and bisphenols blended in CPs and non-additives like heavy metals and persistent organic pollutants such as polychlorinated biphenyls and polycyclic aromatic hydrocarbons adsorbed on CPs [37]. It is also summarized that the observed toxicity of polluted CPs on marine organisms by seven literature reviews. There are much more knowledge accumulated on CPs than on BPs.

BPs are the materials that are biodegradable naturally in the environment because of the action of naturally occurring microorganisms such as bacteria, fungi, and algae. The BPs once entered in environments are also fragmented to smaller pieces: defined as microplastics (<5 mm) and nanoplastics (<1 μm) as shown in Fig. 3. BPs are finally expected to be mineralized to carbon dioxide and water in case of carbohydrate polymers, however, stable dissolved products may be formed during the process. The samples surrounded by the dotted square line in Fig. 3 can be evaluated to determine ecotoxicity. The dissolved sample which may

Table 1 Number of toxicity values used to derive PNEC and EQS of Cybutryne

Test species	Marine species		Freshwater species	
Taxa	Acute test	Chronic test	Acute test	Chronic test
Bacteria	1	1	1	0
Microalgae	10	3	10	4
Macrophyte	4	4	4	5
Zooplankton	19	4	2	1
Benthic invertebrate	3	1	0	0
Fish	4	3	4	2
Total data number	41	16	21	12
Derived PNEC (ng/L)*				
Deterministic	12 (10)	1 (10)	1.2 (10)	0.04 (50)
SSD	11 (10)	2.4 (3)	0.38 (20)	0.5 (3)
Refined SSD	1.5 (20)	1.4 (5)	0.21 (10)	0.22 (5)
Proposed EQS (ng/L)	1.5	1.4	0.21	0.22

*The figures in brackets show assessment factors
Modified original data from Martins et al. Biofouling 34, 34–52, 2018

contain intermediates or final products is applicable for all the ecotoxicity tests, but microplastics (>5 mm) are not suitable for bacterial and algal assay because sample size is much larger than the test organisms. It is possible to identify harmful or non-hazardous fraction and its particle size by evaluating the ecotoxicity of biodegraded water matrices after microfiltration and ultrafiltration.

2.1 Ecotoxicity on Microorganisms and Small Organisms

Manfa et al. [38] reviewed biodegradable polymers and summarized that the degradation of BPs in simulated or real environments is quite investigated and only eleven papers reported their effects on marine organisms, e.g., behavioral and oxidative stress, and potential cascading effects on marine ecosystems. In this section, ecotoxicity assessment results of BPs are outlined for each paper from 14 references containing cited in the review paper. Table 2 indicates the plastics tested (BPs, CPs as control, or commercial products), toxicity test samples (microplastic in seawater, microplastics in diet, solvent extract, or sediment elutriate), test concentrations, exposure type (laboratory or field), test organisms (laboratory or natural environment), and effects observed. Next, the ecotoxicity results from new paper for each test plastic are described.

2.1.1 PLA

Effects of microsized PLA and PBS (53–57 μm) as commercial BPs were assessed on marine microalgae *Chlorella vulgaris* with microsized PE and PA (60–78 μm) as commercial CPs [39]. Algal growth parameters such as inhibition ratio, dry cell weight, pigment contents, and stress indicators such as superoxide dismutase, malondialdehyde, and extracellular polymeric substances were determined. The PLA and PBS inhibited microalgal growth as same as PE and PA did, and PLA among four MPs indicated the strongest inhibitory effect of 48% at 100 mg/l. It was suggested the stress caused by MPs could promote the accumulation of photosynthetic pigments and trigger the evolution of antioxidative substances. The authors suggested co-existing of MPs and microalgae could aggravate the aging of MPs in terms of surface roughness, reduced strength of some functional groups, charge changes, and leaching of toxic chemicals, and all the changes in physicochemical properties could affect the growth of microalgae.

Potential effects of the shredded PLA and PET (200–500 μm) on ascidians were assessed in laboratory test [40]. The 24-h fertilization rates of *Microcosmus exasperatus* were reduced with increasing concentration as 0.1, 0.2, and 0.4 mg/mL of artificial seawater for both shredded particles, but no significant difference was found between materials. Accumulation of particles in adult ascidians cultured in 0, 1, and 10

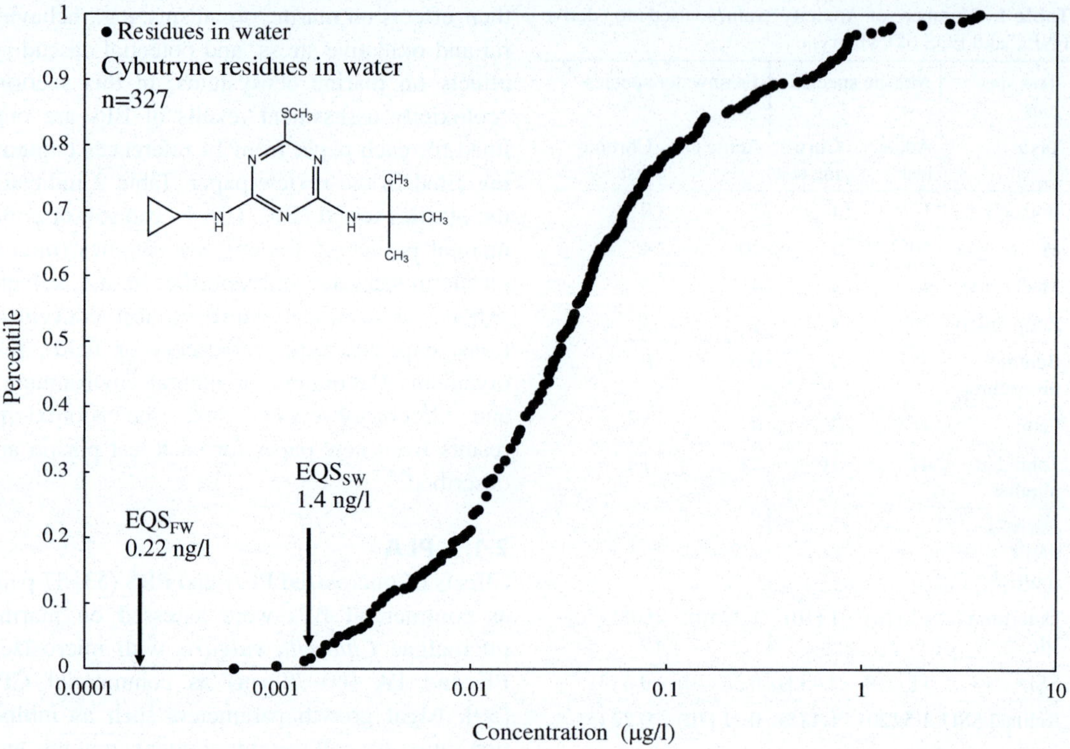

Fig. 2 Environmental risk of cybutryne in water

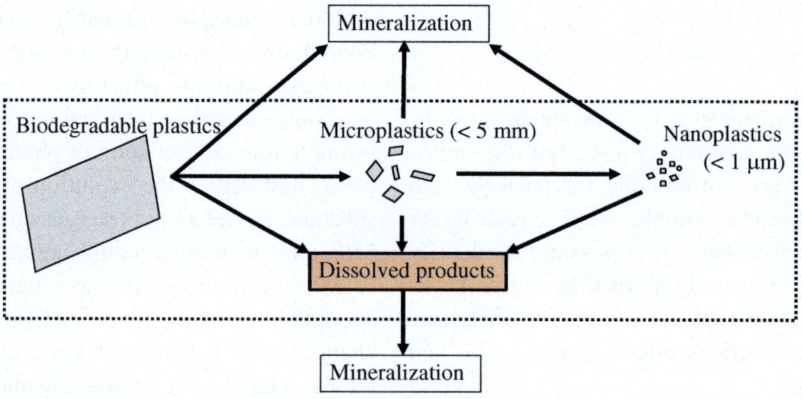

Fig. 3 Biodegradation scheme of biodegradable plastics in seawater

microparticles daily added in the seawater was assessed after 28 days. The average number of particles detected for all the treated ascidians was 9.3+− 4.0, regardless of the materials and the particle numbers (1 or 10) daily added. The PLA and PET plates were deployed in two locations along the Red Sea coast and sessile organisms settled on the test plates were monitored monthly for 3 months. The natural communities after 3 months did not differ significantly between plates depth in any locations, and PET plates submerged exhibited a significant community richness and cover area in comparison to PLA plates.

Blue mussels *Mytilus edulis* were exposed to microsized PLA at two concentrations (10 and 100 μg/L) in artificial seawater (15 L) with aeration for 8 days to evaluate their ecotoxicological effects [41]. The spherical-like PLA powder with its size ranging in 0.8–10 μm was prepared in this study by mechanical milling. No significant mortality, oxidative stress (catalase, glutathione S-transferase, and superoxide dismutase), neurotoxicity (acetylcholinesterase), or immunotoxicity at 8-day exposure were detected. The results showed the lipidome of mussels would be impacted by microplastics at both doses. The authors could not confirm bio-based and compostable particles are safer than petroleum-based ones by comparison of the results of the present study to the literature data on conventional plastics, and suggested that biochemical and metabolomic approaches they took in this study were complementary tools in the environmental risk assessment of microplastics.

Methanol extracts of commercial daily products, covering mostly food contact materials, composed of both BPs and CPs were tested with four kinds of in vitro toxicity assay and high-resolution mass spectrometry. Test samples (3 g) of BPs such as PLA, PHA, PBS, PBAT, starch, cellulose, Bamboo-BioPE composite were extracted with methanol (20 ml) by sonication and the solvent was substituted with DMSO (0.2 ml) by Zimmermann et al. [42]. The DMSO extracts were subjected for baseline toxicity assay with Microtox using bioluminescent bacteria *Allivibrio fischeri*, oxidative stress response assay, and endocrine activity assay for estrogenic and antiandrogenic activities. 67% of total 43 samples induced baseline toxicity, 43% oxidative stress, 23% antiandrogenicity, and one sample estrogenicity. More final products (33 samples) induced toxicity compared to the raw materials (10 samples). The extracts from the CPs (total 30 samples) such as HDPE, LDPE, PS, PP, PET, PVC, and PUR were prepared in the same manner as BPs and were subjected for the toxicity assay [43]. A slightly higher percentage of BPs compared to CPs induced baseline toxicity and antiandrogenicity. The authors identified priority compounds including monomers, oligomers, plastic additives, lubricants, and non-intentionally added substances. They suggested in general that the mean effect strengths of BPs were comparable with CPs, that is, BPs and plant-based materials are similarly toxic as CPs.

Effects of PLA microplastics (66 um) and HDPE (103 μm) on European flat oyster *Osteria edulis* were investigated using outdoor flow-through mesocosm with intact sediment cores in a 5 L of seawater for 2 months [44]. The author added both microplastics to separate cultures of microalgae for 3 days to allow both microplastics to move freely within the container, making the microplastics to have neutral buoyancy. The oyster exposed to high concentration of PLA (80 μg/L) was 2.6 times higher respiration rates than oysters to the high concentrations of HDPE. The filtration rates and shell growth of the oysters did not significantly differ among treatments. This study demonstrated that repeated exposure to high concentrations (80 μg/L) of microplastics in water column could alter assemblages in marine habitat by reducing the abundance of benthic fauna.

Effects of microplastics of PLA (236 μm), HDPE (103 μm), and PVC (131 μm) on lugworm *Arenicola marina* in sediments and the primary productivity of the sediments were investigated [45]. Respiration rates of lugworms increased, but microalgal biomass of surface sediment decreased at high concentrations of microplastics and responses were strongest to PVC. The authors concluded both the CPs and BPs tested in sandy sediments affected the health and behavior of lugworms and directly or indirectly reduced primary productivity.

Taylor et al. [46] investigated the toxicity of bioabsorbable polymers, used for alternative to metal implants for internal fracture fixation, on photobacteria *Photobacterium phosphoreum* using a Microtox, which could quantify toxicity as reduction of bacterial light emission. The test polymers, PLA, PCL, PGA, POE, and PHBV undergo hydrolysis and produce degradation product components such as lactic acid, hydroxy caproic acid, glycolic acid, intermediate products (pentaerythritol dipropionate,

Table 2 Ecotoxicological assessment of biodegradable plastics by laboratory/field experiments

Biodegradable plastics (BPs)	Conventional plastics (CPs) or reference material	Toxicity test samples	Test concentrations	Laboratory/Field exposure	Test organism, exposure	Effects observed	References
PLA (mean 57 μm), PBS (mean 53 μm)	PE (mean 78 μm), PA (mean 60 μm)	MPs in seawater	0, 10, 100, 1000 mg/L (seawater)	Laboratory	*Chlorella vulgaris* (microalgae, marine), 11 days	Algal growth inhibition Stimulation in pigment content and extracellular polymeric substances	[39]
PLA (200–500 μm) PLA plates	PET (200–500 μm) PET plates	MPs in seawater Plate in seawater	0, 0.1, 0.2, 0.4 mg/mL (seawater) 1, 10 particles daily (seawater) Plates submerged in Red Sea	Laboratory/Field	*Microcosmus exasperates* (ascidian), 24 h *Microcosmus exasperates* (ascidian), 28 days Natural community assemblage, 3 months	Reduced fertilization rate Accumulation in adult tissues Higher community richness in PET plates than PLA plates	[40]
PLA (0.8–10 μm)	HDPE, PS, PE	MPs in seawater	0, 10, 100 μg/L (seawater)	Laboratory	*Mytilus edulis* (blue mussel, marine), 8 days	No significant oxidative stress, neurotoxicity, or immunotoxicity	[41]
PLA, PHA, PBS, PBAT, starch, Cellulose, Bamboo-BioPE composite	HDPE, LDPE, PS, PP, PET, PVC, PUR	Methanol extracts of composite products and raw materials	0, 0.2, 0.5, 1.0% (v/v) Extracts substituted with dimethyl sulfoxide (0.2 ml), prepared from 3 g materials	Laboratory	Baseline toxicity by Microtox using *Aliivibrio fischeri* (bacteria), 30 min-EC20 Oxidative stress by AREc32 assay Endocrine activity by YES and YAAS assay	Reduced bioluminescence Altered oxidative stress response Induced estrogenic activity and anti-androgenic activity	[42, 43]

(continued)

Table 2 (continued)

Biodegradable plastics (BPs)	Conventional plastics (CPs) or reference material	Toxicity test samples	Test concentrations	Laboratory/Field exposure	Test organism, exposure	Effects observed	Referen-ces
PLA (0.6–363 μm, mean 66 μm)	HDPE (0.5–316 μm, mean 103 μm)	MPs in seawater	0, 0.8, 80 μg/L (seawater) dosed repeatedly for 60 days	Outdoor mesocosm	*Ostrea edulis* (oyster, marine) with intact sediment cores and seawater, 60 days	Higher oyster respiration rate at high dose PLA than high dose HDPE Differred benthic assemblage structures and reduced species richness and the total number of organisms by repeated high dose PLE and HDPE	[44]
PLA (1.4–707 μm, mean 236 μm)	HDPE (2.5–316 μm, mean 103 μm), PVC (8.7–478 μm, mean 131 μm)	MPs in sediment	0, 0.02, 0.2, 2% (wet sediment weight)	Outdoor mesocosm	*Arenicola marina* (lugworm, marine), outdoor mesocosm with wet sieved sand and seawater, 31 days	Increased lugworm respiration rate by high dose of 3 MPs Reduced diatom biomass and chlorophyll-a contents of surface sediments by 3 MPs	[45]
PLA, PCL, PGA, POE, PHBV film and their degradation products	No	Water incubated with BPs film	Quartered films from 1-inch disk incubated in waters up to 16 weeks	Laboratory	*Photobacterium phosphoreum* same as *Aliivibrio fischeri* (bacteria), 5 min-EC50	Reduction of bioluminescence	[46]
PHB (10–90 μm)	PE (10–90 μm)	MPs in seawater	1000 MPs/mL	Laboratory	*Mytilus edulis* (mussel, marine), 96 h	Altered oxidative stress biomarkers	[47]
PHB (<80 mesh)	No	MPs in diet	0, 2% (PHB supplemented diet)	Laboratory	*Liza haemotocheila* (fish, marine), 28 days	Higher growth rate and weight gain Higher antioxidant enzyme activity Higher immune gene expression	[48]

(continued)

Table 2 (continued)

Biodegradable plastics (BPs)	Conventional plastics (CPs) or reference material	Toxicity test samples	Test concentrations	Laboratory/Field exposure	Test organism, exposure	Effects observed	References
PHB	No	MPs dissolved in chloroform were sprayed onto feed pellets	0, 1, 3, 5% (PHB supplemented diet)	Laboratory	*Litopenaeus vannamei* (pacific white shrimp), 35 days	Beneficial effect on intestine microbiota	[49]
Commercial products							
Mater-Bi films consisting of starch, bio-PE, and a natural plasticizer	Film made of microcrystalline cellulose	Sediment elutriates with film after 6 and 12 months	100% + diluitions 100% + diluitions 25% vol/vol of elutriate	Laboratory	*Dunaliella tertiolecta* (algae, marine), 72 h *Paracentrotus lividus* (sea urchin, marine), 48 h *Dicentrarchus labrax* (fish, marine), 14 days	No significant inhibition on algal growth Low/absent embryo-toxicity for sea urchin No lipid peroxidation and micronuclei alteration	[50]
Mater-Bi bags consisting of vegetable oils and corn starch	Cellulose filter paper	Bags (14 cm × 14 cm, 20 μm thickness) buried in sediments in outdoor tank with seawater		Outdoor mesocosm	*Cymodocea nodosa* and *Zostera noltei* (seagrasses, marine), 180 days	Increased plant intra/interspecific-competition Altered sediment geochemistry	[51]
Carrier bags made of corn starch	Commercial carrier bags made of HDPE	Bags (38 × 46 cm, 65 μm thickness)	Bags positioned flat using metal pins on pristine mud-flat, 9 weeks	Field	Benthic infaunal invertebrates, 9 weeks	Anoxic conditions created by BPs/CPs within the sediment Reduced primary productivity, organic matter, and infaunal abundance	[52]

trans-cyclohexane dimethanol, 1,6-hexane diol, pentaerythritol, propionic acid), and both butyric acid and valeric acid, respectively. The 5 min EC50 of the degradation products were ranging from 72,000 to 336,000 mg/L, which concentrations are too high to cause toxicity in environment. Only PGA film among other specimens in pure water or Tris buffer solution after 16 weeks of incubation degraded by 63–65% of the initial weight and the incubation solutions completely inhibited light emission. It seems the solution pH (2.5–2.9) lowered because glycol acid produced by hydrolysis occurred the reduction of light emission.

2.1.2 PHB

Effects of PHB and PE microplastics (10–90 μm) on detoxifying enzymes in digestive glands and gills of *Mytilis edulis* were investigated with 4-day laboratory experiments [47]. The blue mussels were exposed to test specimen ($2*10^5$ particles/200 ml) in seawater for 4 days and the gill and digestive gland were subjected for determination of enzyme activity. They aimed to compare the effects of MPs administered alone or in combination with fluoranthene on detoxifying enzyme activities, however, the results omitting the combination effects with fluoranthene are compared to the controls here. In the gills, catalase (CAT), glutathione peroxidase (GPx), glutathione S-transferase (GST) activity levels occurred by PHB were significantly lower than the ones occurred by PE. In digestive glands, PHB treatment reduced superoxide dismutase, CAT, and GPx activity levels than PE treatment, but activated GST levels. Thus, PHB results in altered levels of some oxidative stress biomarkers similar to oil-based PE.

Poly-beta-hydroxybutyrate (PHB) is a natural polymer of the short chain fatty acid beta-hydroxybutyrate, which acts as a microbial control agent. PHB, as a dietary supplement, has been reported to promote growth in fish and shellfish such as tilapia, European sea bass, Siberian sturgeon, Chinese mitten crab, and sea cumber [48]. Effects of dietary PHB on the composition of intestine microbiota, and the mammalian target of the rapamycin (mTOR) signaling-related gene expression in Pacific white shrimp *Litopenaeus vannamei* fed diets containing different levels of PHB for 35 days. Effects of PHB on the growth, immune-related enzyme activity, and transcriptome-based gene expression in fish *Liza haematocheila*, were investigated [48]. PHB dietary supplementation (2%) for 28 days could increase fish growth, weight gain, and the antioxidant enzyme activity including total antioxidant capacity, catalase, and superoxide dismutase, and activate MAPK/PI3K-Akt/TNF/NF-κB/TCR/TLR signaling in response to PHB stimulation. The authors suggested that immuno-enhancement of PHB might be more related to the antioxidant system based on both enzyme activities and transcriptome-based signaling pathways.

Effects of dietary PHB on microbiota composition and mammalian target of the rapamycin (mTOR) signaling-related gene expression in the intestines of the Pacific white shrimp *Litopenaeus vannamei* fed diets containing different levels of PHB 0, 1, 3, 5%(w/w) for 35 days [49]. Dietary PHB altered the composition and diversity of intestine microbiota and the diversity suppressed by PHB with increasing order of the contents. The relative mRNA expression levels of the mTOR signaling-related genes increased by PHB. The results suggested dietary PHB supplementation had beneficial effects on the intenstine health of *L.vannamei* by modulating the composition of intestine microbiota and activating mTOR signaling.

2.1.3 Commercial Products

The commercial bags, the Mater-Bi grade HF03V1, made of starch, biodegradable polyesters and a natural plasticizer, are biodegradable and disintegrable under composting conditions [50]. The bags were powdered by cryogenic grinding and the pieces of the film (4 cm × 4 cm) buried in the marine sediments of the reactors were monitored the progress of disintegration for up to 12 months. The sediments (1 kg dry) collected at 6 and 12-month exposure were stirred with synthetic marine water (3 L) for 24 h at 4 centigrade, and the liquid phase (elutriate) filtered was subjected to a battery of standardized bioassays using algae, sea urchin, and fish. There was no significant algal growth inhibition, low or absent embryo toxicity to sea urchin, and no significant lipid peroxidation and micronuclei alteration of the fish erythrocytes.

Effects of biodegradable plastic bags on seagrass meadows inhabitating sandy bottoms at coastal zones were assessed by mesocosm experiments using outdoor tanks (7000 L) with a continuous flow of natural seawater [51]. A sample piece (14 cm × 14 cm, 20 μm thickness) of Mater-Bi bag consisting of vegetable oils and corn starch or a cellulose filter paper as the control was buried in the top of the sediments. After 6 months of sediment exposure, the bag retained 85% of the initial weight, the biodegradable Mater-Bi bags degrade slowly, and reduced sediment pore water oxygen concentration and pH. Seagrass *Cymodocea nodosa* root spread and vegetative recruitment increased compared to control. The authors suggested the biodegradable bags altering sediment geochemistry could stimulate the spatial segregation of seagrass clones and influence coexistence.

A field experiment was performed to assess the impacts of disintegrated plastic bags on marine assemblages and ecosystem functioning at Booterstown Marsh, Ireland [52]. The commercial carrier bags made of corn starch and HDPE (38 × 46 cm, 65 μm thickness) were randomly interspersed, by ten replications, on the surface in an area (10 m × 25 m) of pristine mud-flat, with positioned flat using metal pins at each corner. After 9 weeks, the presence of either type of bags occurred in anoxic conditions within the sediment along with diminished primary productivity and organic matter and significantly lower abundances of infaunal invertebrates. The authors suggested that both BPs and CPs bags can rapidly change marine assemblages and the ecosystem services.

2.2 Effects of Chemicals Blended in Plastics on Fish

Phthalates (phthalate esters, PAEs) are used in the production of a wide range of consumer products including plastic toys, wallpaper, cosmetics, fragrances, and shampoos. However, there are several reports that PAEs are harmful to fish. For example, Qian et al. [53] have reported that PAEs induce transcriptional alterations of genes related to spinal development, leading to spinal birth defects and abnormal behavior in zebrafish embryo. Pu et al. [54] have reported that di-(2-ethylhexyl) phthalate (DEHP) and di-*n*-octyl phthalate induce minor defects in the developmental morphology of zebrafish larvae. PAEs have also been reported to have endocrine disrupting effects.

Endocrine hormones play an important role in maintaining homeostasis because in very small amounts they are able to regulate physiological actions. These hormones are made in the endocrine glands, secreted into the blood, and then transported to target organs where they bind to their target receptor. Some chemicals are able to disrupt endocrine function by binding to target receptors and acting as hormones, or by directly interfering with hormonal function. Such chemicals are called endocrine disruptors.

In zebrafish, mono-(2-ethylhexyl) phthalate (MEHP) disrupts the production of sex hormones by altering the expression of key genes related to the steroidogenic pathways of vitellogenin, follicle-stimulating hormone β, luteinizing hormone β, and cytochrome P450 aromatase A [55]. MEHP also disrupts the thyroid endocrine system by altering the expression of the genes encoding thyroid stimulating hormone β, thyronine deiodinase 1, thyronine deiodinase 2, and transthyretin in zebrafish [56]. Similarly, DEHP disrupts the thyroid endocrine system by altering the expression of genes encoding thyroid stimulating hormone β, thyronine deiodinase 2, and transthyretin [57]. Recently, it has been reported that DEHP also suppresses swimming performance in zebrafish [58].

Bisphenol A (BPA) is a very important raw material in the plastics industry, serving as a building block for the production of polycarbonate plastics, which are frequently used to line food and beverage containers, or as an additive in other plastics. BPA is also a component of the epoxy resins that are used in some dental materials [59]. Grand View Research reported that the global demand for BPA was around 7000 kilotons in 2013, and in 2020 it reached around 9600 kilotons, growing at a compound annual growth rate of 4.7% from 2014 to 2020. There are many reports that BPA has endocrine-disrupting potential in teleost fish. For example, BPA is reported to suppress reproduction in zebrafish [60], medaka [61], and fathead minnow [62], disrupt gonadal sex differentiation and induce testis–ova or male-to-female sex reversal in zebrafish [60] and medaka [63], and increase vitellogenin expression in male zebrafish [64], mangrove rivulus (*Kryptolebias marmoratus*, [65], goldfish (*Carassius auratus*, [66], medaka [67], and fathead minnow [68]. Together, these reports indicate that BPA has estrogenic potential in teleost fish.

2.3 Effects of Microplastics on Fish

Plastic products are indispensable in modern society. About 322 million tons of plastic were produced in 2015; about half of this is garbage, and at least 8 million tons is discharged into the sea every year (from the Ministry of the Environment). Microplastics (MPs) are plastic particles with a length of 5 mm or less. Primary microplastics are manufactured already 5 mm or less in length and include microfibers and microbeads. Secondary microplastics are particles that are shed from larger plastic products as they are degraded by weathering processes. MPs have been identified in oceans and freshwater habitats around the world [69], reviewed by Yang et al. [70]. In addition, MPs of various colors and transparencies have been found in the ocean [71] and inside of commercial fish [72].

Two important factors for estimating the harmful effects of MPs on fish are retention time and excretion time, which allow for assessment of the extent to which hazardous chemical substances adsorbed onto MPs are absorbed into the body after ingestion of MPs. Studies in red sea bream and mummichog have shown that more than 95% of MPs are excreted from the body within 25 h of ingestion [73]. However, Okamoto et al. [74] have reported in two freshwater fish (Japanese medaka and zebrafish) and two marine fish (Indian medaka and clown anemonefish) that MP excretion times vary widely among fish of the same species, and that some individuals still had MP particles remaining in their gastrointestinal tracts at more than 24 h after exposure.

In 2020, Jacob et al. [75] published a review discussing the ecotoxicological influence of MPs on fish, concluding that MPs impact fish metabolism, fitness, immune system, behavior, sensory system, neuromuscular function, alimentary and excretory systems, microbiome, and circulatory and respiratory systems. Recently, it has been reported that MPs and nanoplastics exert combination effects with chemical substances. He et al. [76] have reported that MPs and nanoplastics enhance the toxicity of triphenyl phosphate, a plasticizer, and fire retardant, in zebrafish. Wang et al. [77] have reported that polystyrene MPs increase the estrogenic effects of 17α-ethinylestradiol, a synthetic estrogen used in oral contraceptive pills, in male marine medaka (*O. melastigma*). Takai et al. [78] have reported that simultaneous exposure to the antianxiety agent diazepam and polystyrene microplastics suppresses medaka social behavior. However, Lu et al. [79] have reported an antagonistic effect between polystyrene and the antibiotic sulfamethoxazole, which slightly decreased their combined toxicity. The implications here are that combinations of MPs and chemical substances can have synergistic or antagonistic effects and that more studies are needed to fully understand the harmful effects of plastic–chemical combinations.

3 Ecotoxicological Concerns on Biodegradable Plastics

If the ecotoxicity of a sample is evaluated using non-standardized toxicity tests, it may be difficult to compare both test results qualitatively and quantitatively. Therefore, ecotoxicity of the degradation products in the biodegradation process from plastics tested should be evaluated using internationally standardized toxicity tests with risk assessment methodology. According to the draft document on marine ecotoxicity testing scheme for decomposition intermediates from biodegradable plastic materials in marine environments [80], ecotoxicological effects on marine organisms shall be determined by comparing marine matrices from preceding marine biodegradation tests produced with or without the addition of a test material. Although the draft has not yet been finalized and has not yet been voted on, the proposed tests at this time are composed of (1) marine algal growth inhibition test modified ISO 10253, (2) marine copepods test modified ISO 14669, (3) fish acute test modified OECD 203, and (4) luminescent bacteria test modified ISO 11348, and (1), (2), and (4) tests are mandatory and (3) fish test is optional. The test species with exposure time in parentheses are specified as *Skeletonema costatum* or *Phaeodactylum tricornutum* as marine algae (72 h exposure), *Acartia tonsa*, *Tisbe battaglial*, or *Nitocra spinipes* as marine copepods (48 h exposure), *Cyprinodon variegatus*, *Dicentrachus labrax*, or *Pagrus major* as fish (96 h exposure), and *Vibrio fisheri* as luminescent bacteria (30 min exposure). Thus, the ecotoxicity of intermediates decomposed from BPs in water matrices should be evaluated with the standardized test methods.

It is important to clarify whether the harmful effects of microplastics are caused by dissolved components produced during biodegradation or by the nanoplastics produced by further decomposition. Several manufactured nanoplastics such as polycarbonate, poly(methyl methacrylate), and latex have induced a variety of effects on growth, development, behavior, reproduction, and mortality of several organisms and trophic transfer of nanoplastics have been also investigated [81, 82]. Thus, the evidence on the effects of primary nanoparticles, especially on marine organisms, has been accumulated, however, the studies are limited to primary nanoplastics synthesized in the laboratories. Since the effect of secondary nanoplastics produced by degradation processes in the environments has not been assessed, the effects of biodegradable nanoplastics on aquatic organisms are completely unknown. Gonzalez-Pleiter et al. [83] reported secondary PHB nanoplastics were released from PHB microplastics by abiotic degradation under environmentally representative conditions and they were harmful for three freshwater organisms (cyanobacteria *Anabaena* sp., green algae *Chlamydomonas reinhardtii*, and crustacean *Daphnia magna*) that play a key role in freshwater ecosystems. The authors suggested by all results put together that biodegradable plastic does not mean safe for the environment in the case of PHB.

Another concern is the hazards associated with chemical additives within micro(nano)plastics [37, 84]. Plastic polymers are rarely used in their pristine form and incorporate chemical additives to give specific functions to the material [85]. Chemical additives are on the basis of key functional properties as processability improvers like plasticizers, surface protectors/modifiers like antistatic agents, material protectants like antioxidants, physical–chemical property improvers like flame retardants, and functionalizing agents like coloring agents [86]. The additives are rarely chemically bound to plastic polymers and have potential to leach from plastic to water as degrading. Although the ecotoxicity of some additives known as persistent, bioaccumulative, and toxic (PBT) compounds has been evaluated, there are concerns about the ecological risks they present as a component of plastics containing the additives. Some additives have been reported to be higher hazardous than biocides [84]. Additives blended in CPs and non-additives adsorbed on CPs have been reported by literature review [37]. Since the BPs developed also contains the same chemical additives as CP, it is necessary to use less hazard additives.

Based on above-mentioned methodology, we are making efforts to evaluate environmental toxicity of newly developed bionylons introduced in Chap. 2.

References

1. OECD, Manual for investigation of HPV chemicals. Chapter 4: Initial assessment of data. 4.2 Guidance for the Initial Assessment of Aquatic Effects (2002). https://www.oecd.org/chemicalsafety/risk-assessment/2483645.pdf
2. ISO, Ships and marine technology—Risk assessment on anti-fouling systems on ships—Part 1: Marine environmental risk assessment method of biocidally active substances used for anti-fouling systems on ships. ISO 13073-1 (2012)
3. ISO, Water quality—Determination of the inhibitory effect of water samples on the light emission of *Vibrio fischeri* (Luminescent bacteria test)—Part 1: Method using freshly prepared bacteria. ISO 11348-1 (2007a)
4. ISO, Water quality—Determination of the inhibitory effect of water samples on the light emission of *Vibrio fischeri* (Luminescent bacteria test)—Part 2: Method using liquid-dried bacteria. ISO 11348–2 (2007b)
5. ISO, Water quality—Determination of the inhibitory effect of water samples on the light emission of *Vibrio fischeri* (Luminescent bacteria test)—Part 3: Method using freeze-dried bacteria. ISO 11348–3 (2007c)
6. ISO, Water quality—Marine algal growth inhibition test with *Skeletonema* sp. and *Phaeodactylum tricornutum.* ISO 10253 (2016a)
7. ISO, Water quality—Growth inhibition test with the marine and brackish water macroalga *Ceramium tenuicorne.* ISO 10710 (2010)
8. ISO, Water quality—Determination of acute lethal toxicity to marine copepods (Copepoda, Crustacea) ISO 14669 (1999)
9. ASTM, Standard Guide for Conducting Life-Cycle Toxicity Tests with Saltwater Mysids. ASTM E1191-03a (2014a)
10. ASTM, Standard Guide for Conducting Short-Term Chronic Toxicity Tests with Echinoid Embryos. ASTM E1563-21a (2022a)
11. ISO, Water quality - Determination of the acute toxicity to the marine rotifer *Brachionus plicatilis.* ISO 19820 (2016b)
12. ASTM, Standard Guide for Conducting Acute, Chronic, and Life-Cycle Aquatic Toxicity Tests with Polychaetous Annelids. ASTM E1562-22 (2022b)
13. ASTM, Standard Guide for Conducting Static Short-Term Chronic Toxicity Tests Starting with Embryos of Four Species of Saltwater Bivalve Molluscs. ASTM E724-21 (2021)
14. ISO, Water quality—Determination of the toxicity of water samples on the embryo-larval development of Japanese oyster (*Crassostrea gigas*) and mussel (*Mytilus edulis* or *Mytilus galloprovincialis*). ISO 17244 (2015)
15. ASTM, Standard Test Method for Measuring the Toxicity of Sediment-Associated Contaminants with Estuarine and Marine Invertebrates. ASTM E1367-03 (2014b)
16. OECD, Guidelines for the Testing of Chemicals, Test No. 203: Fish, Acute Toxicity Test. OECD Publishing (2019a)
17. OECD, Guidelines for the Testing of Chemicals, Test No. 212: Fish, Short-term Toxicity Test on Embryo and Sac-Fry Stages. OECD Publishing (1998)
18. OECD, Guidelines for the Testing of Chemicals, Test No. 210: Fish, Early-life Stage Toxicity Test. OECD Publishing (2013a)
19. OECD, Guidelines for the Testing of Chemicals, Test No. 234: Fish Sexual Development Test. OECD Publishing (2011)
20. OECD, Guidelines for the Testing of Chemicals, Test No. 229: Fish Short Term Reproduction Assay. OECD Publishing (2012)
21. OECD, Guidelines for the Testing of Chemicals, Test No. 240: Medaka Extended One Generation Reproduction Test (MEOGRT). OECD Publishing (2015)
22. EU, Directive 2010/63/EU of the European Parliament and Council of 22 September on the protection of animals used for scientific purposes. Off J. Eur. Union **L276**, 33–79 (2010)
23. OECD, Guidelines for the Testing of Chemicals, Test No. 236: Fish Embryo Acute Toxicity (FET) Test. OECD Publishing (2013b)
24. S.E. Martins, G. Fillmann, A. Lillicrap, K.V. Thomas, Review: ecotoxicity of organic and organometallic antifouling co-biocides and implications for environmental hazard and risk assessments in aquatic ecosystems. Biofouling **34**(1), 34–52 (2018). https://doi.org/10.1080/08927014.2017.1404036
25. H.R. Ali, M.M. Arifin, M.A. Sheikh, N.A.M. Shazili, Z. Bachok, Occurrence and distribution of antifouling biocide Irgarol-1051 in coastal waters of peninsular Malaysia. Mar. Pollut. Bull. **70**, 253–257 (2013). https://doi.org/10.1016/j.marpolbul.2013.02.024
26. C. Basheer, K.S. Tan, H.K. Lee, Organotin and Irgarol-1051 contamination in Singapore coastal waters. Mar. Pollut. Bull. **44**, 697–703 (2002). https://doi.org/10.1016/S0025-326X(01)00330-7
27. K. Carbery, R. Owen, T. Frickers, E. Otero, J. Readman, Contamination of Caribbean coastal waters by the antifouling herbicide Irgarol 1051. Mar. Pollut. Bull. **52**, 635–644 (2006). https://doi.org/10.1016/j.marpolbul.2005.10.013
28. L.W. Hall, W.D. Killen, R.D. Anderson, R. Balcomb, P. Gardinali, Ecological risk of Irgarol 1051 and its major metabolite in coastal California marinas and reference arease. Mar. Pollut. Bull. **58**, 702–710 (2009). https://doi.org/10.1016/j.marpolbul.2008.12.019
29. H. Harino, Y. Yamamoto, S. Eguchi, S. Kawai, Y. Kurokawa, T. Arai, M. Ohji, H. Okamura, N. Miyazaki, Concentrations of antifouling biocides

in sediment and mussel samples collected from Otsuchi Bay, Japan. Arch. Environ. Contam. Toxicol. **52**, 179–188 (2007). https://link.springer.com/article/1007/s00244-006-0087-2
30. N.S. Kim, S.H. Hong, J.G. An, K.H. Shin, W.J. Shim, Distribution of butyltins and alternative antifouling biocides in sediments from shipping and shipbuilding areas in South Korea. Mar. Pollut. Bull. **95**, 484–490 (2015). https://doi.org/10.1016/j.marpolbul.2015.03.010
31. S. Knutson, C.A. Downs, R.H. Richmond, Concentrations of Irgarol in selected marinas of Oahu, Hawaii and effects on settlement of coral larval. Ecotoxicology **21**, 1–8. (2012). https://link.springer.com/article/10.1007/s10646-011-0752-8
32. H. Okamura, I. Aoyama, Y. Ono, T. Nishida, Antifouling herbicides in the coastal waters of western Japan. Mar. Pollut. Bull. **47**, 59–67 (2003). https://doi.org/10.1016/S0025-326X(02)00418-6
33. R. Owen, A. Knap, M. Toaspern, K. Carbery, Inhibition of coral photosynthesis by the antifouling herbicide Irgarol 1051. Mar. Pollut. Bull. **44**, 623–632 (2002). https://doi.org/10.1016/S0025-326X(01)00303-4
34. A. Saleh, S. Molaei, N.S. Fumani, E. Abedi, Antifouling paint booster biocides (Irgarol 1051 and diuron) in marinas and ports of Bushehr, Persian Gulf. Mar. Pollut. Bull. **105**, 367–372 (2016). https://doi.org/10.1016/j.marpolbul.2016.02.037
35. Y. Shapozhnikova, E. Wirth, K. Schiff, J. Brown, M. Fulton, Antifouling pesticides in the coastal waters of Southern California. Mar. Pollut. Bull. **541**, 1972–1978 (2007). https://doi.org/10.1016/j.marpolbul.2007.09.026
36. M.A. Sheikh, F.S. Juma, P. Staehr, K. Dahl, R.J. Rashid, M.S. Mohammed, A.M. Ussi, H.R. Ali, Occurrence and distribution of antifouling biocide Irgarol-1051 in coral reef ecosystems, Zanzibar. Mar. Pollut. Bull. **109**, 586–590 (2016). https://doi.org/10.1016/j.marpolbul.2016.05.035
37. J. Palmer, S. Herat, Ecotoxicity of microplastic pollutants to marine organisms: a systematic review. Water Air Soil Pollut. **232**, 195 (2021). https://doi.org/10.1007/s11270-021-05155-7
38. L. Manfa, V. Marengo, G. Libralato, M. Costantini, F. De Falco, M. Cocca, Biodegradable polymers: a real opportunity to solve marine plastic pollution. J. Hazard Mater. **416**, 125763 (2021). https://doi.org/10.1016/j.jhazmat.2021.125763
39. Y.Y. Su, Z.R. Chen, Y.P. Hou, S.Y. Lin, L. Gao, Z.Z. Wang, R.Q. Bao, L.C. Peng, Biodegradable and conventional microplastics posed similar toxicity to marine algae *Chlorella vulgaris*. Aquat. Toxicol. **244**, 106097 (2022). https://doi.org/10.1016/j.aquatox.2022.106097
40. G. Anderson, N. Shenkar, Potential effects of biodegradable single-use items in the sea: polylactic acid (PLA) and solitary ascidians. Environ. Pollut. **268**, 115364 (2021). https://doi.org/10.1016/j.envpol.2020.115364
41. A. Khalid, A. Zalouk-Vergnoux, S. Benali, R. Mincheva, J.M. Raquez, S. Bertrand, L. Poirier, Are bio-based and biodegradable microplastics impacting for blue mussel (*Mytilus edulis)?*. Mar. Pollut. Bull. **167**, 112295 (2021). https://doi.org/10.1016/j.marpolbul.2021.112295
42. L. Zimmermann, A. Dombrowski, V¨olker C, Wagner M, Are bioplastics and plant-based materials safer than conventional plastics? In vitro toxicity and chemical composition. Environ. Int. **145**, 106066 (2020). https://doi.org/10.1016/j.envint.2020.106066
43. L. Zimmerman, G. Dierkes, T.A. Ternes, C. Volker, M. Wagner, Benchmarking the invitro toxicity and chemical composition of plastic consumer products. Environ. Sci. Technol. **53**, 11467–11477 (2019). https://doi.org/10.1021/acs.est.9b02293
44. D.S. Green, Effects of microplastics on European flat oysters, Ostrea edulis and their associated benthic communities. Environ. Pollut. **216**, 95–103 (2016). https://doi.org/10.1016/j.envpol.2016.05.043
45. D.S. Green, B. Boots, J. Sigwart, S. Jiang, C. Roch, Effects of conventional and biodegradable microplastics on a marine ecosystem engineer (Arenicola marina) and sediment nutrient cycling. Environ. Pollut. **208**, 426–434 (2016). https://doi.org/10.1016/j.envpol.2015.10.010
46. M.S. Taylor, S.U. Daniels, K.P. Andriano, J. Heller, Six bioabsorbable polymers: in vitro acute toxicity of accumulated degradation products. J. Appl. Biomater. **5**(151), I57 (1994). https://doi.org/10.1002/jab.770050208
47. G. Magara, F.R. Khan, M. Pinti, K. Syberg, A. Inzirillo, A.C. Elia, Effects of combined exposures of fluoranthene and polyethylene or polyhydroxybutyrate microplastics on oxidative stress biomarkers in the blue mussel (Mytilus edulis). J. Toxicol. Environ. Health A **82**(10), 616–625 (2019). https://doi.org/10.1080/15287394.2019.1633451
48. G. Qiao, Q. Sun, M. Zhang, C. Xu, T. Lv, Z. Qi, W. Yang, Q. Li, Antioxidant system of soiny mullet (Liza haematocheila) is responsive to T dietary poly- β-hydroxybutyrate (PHB) supplementation based on immune-related enzyme activity and de novo transcriptome analysis. Fish Shellfish Immunol. **95**, 314–327 (2019). https://doi.org/10.1016/j.fsi.2019.10.042
49. Y. Duan, Y. Zhang, H. Dong, Y. Wang, J. Zhang, Effects of dietary poly- β-hydroxybutyrate (PHB) on microbiota composition and the mTOR signaling pathway in the intestines of Litopenaeus vannamei. J. Microbiol. **55**(12), 946–954 (2017). https://doi.org/10.1007/s12275-017-7273-y
50. T. Campani, S. Casini, I. Caliani, C. Pretti, M.C. Fossi, Ecotoxicological investigation in three model species exposed to elutriates of marine sediments inoculated with bioplastics. Front. Mar. Sci. **7**, 229 (2020). https://doi.org/10.3389/fmars.2020.00229
51. E. Balestri, V. Menicagli, F. Vallerini, C. Lardicci, Biodegradable plastic bags on the seafloor: a future

threat for seagrass meadows? Sci. Total Environ. **605–606**, 755–763 (2017). https://doi.org/10.1016/j.scitotenv.2017.06.249
52. D.S. Green, B. Boots, D.J. Blockley, C. Rocha, R. Thompson, Impacts of discarded plastic bags on marine assemblages and ecosystem functioning. Environ. Sci. Technol. **49**, 5380–5389 (2015). https://doi.org/10.1021/acs.est.5b00277
53. L. Qian, J. Liu, Z. Lin, X. Chen, L. Yuan, G. Shen, W. Yang, D. Wang, Y. Huang, S. Pang, X. Mu, C. Wang, Y. Li, Evaluation of the spinal effects of phthalates in a zebrafish embryo assay. Chemosphere **249**, 126144 (2020). https://doi.org/10.1016/j.chemosphere.2020.126144
54. S.Y. Pu, N. Hamid, Y.W. Ren, D.S. Pei, Effects of phthalate acid esters on zebrafish larvae: Development and skeletal morphogenesis. Chemosphere **246**, 125808 (2020). https://doi.org/10.1016/j.chemosphere.2019.125808
55. Y. Zhu, R. Hua, Y. Zhou, H. Li, S. Quan, Y. Yu, Chronic exposure to mono-(2-ethylhexyl)-phthalate causes endocrine disruption and reproductive dysfunction in zebrafish. Environ. Toxicol. Chem. **35**(8), 2117–2124 (2016). https://doi.org/10.1002/etc.3369
56. W. Zhai, Z. Huang, L. Chen, C. Feng, B. Li, T. Li, Thyroid endocrine disruption in zebrafish larvae after exposure to mono-(2-ethylhexyl) phthalate (MEHP). PLoS ONE **9**(3), e92465 (2014). https://doi.org/10.1371/journal.pone.0092465
57. P.P. Jia, Y.B. Ma, C.J. Lu, Z. Mirza, W. Zhang, Y.F. Jia, W.G. Li, D.S. Pei, The effects of disturbance on Hypothalamus-Pituitary-Thyroid (HPT) axis in Zebrafish Larvae after exposure to DEHP. PLoS ONE **11**(5), e0155762 (2016). https://doi.org/10.1371/journal.pone.0155762
58. W. Huang, J. Xiao, X. Shi, S. Zheng, H. Li, C. Liu, K. Wu, Effects of di-(2-ethylhexyl) phthalate (DEHP) on behavior and dopamine signaling in zebrafish (*Danio rerio*). Environ. Toxicol. Pharmacol. **93**, 103885 (2022). https://doi.org/10.1016/j.etap.2022.103885
59. M.D. Shelby, NTP-CERHR monograph on the potential human reproductive and developmental effects of bisphenol A. National Toxicology Program, Center For The Evaluation of Risks To Human Reproduction, MON,22, v, vii-ix, 1–64 passim (2008)
60. J. Chen, K.S. Sail, Y. Liu, L. Li, Y. Zhao, Y. Jia, C. Bai, R.L. Tanguay, Q. Dong, C. Huang, Developmental bisphenol A exposure impairs sperm function and reproduction in zebrafish. Chemosphere **169**, 262–270 (2017). https://doi.org/10.1016/j.chemosphere.2016.11.089
61. Y. Kawashima, Y. Onishi, N. Tatarazako, H. Yamamoto, M. Koshio, T. Oka, Y. Horie, H. Watanabe, T. Nakamoto, J. Yamamoto, H. Ishikawa, T. Sato, K. Yamazaki, T. Iguchi, Summary of 17 chemicals evaluated by OECD TG229 using Japanese medaka, *Oryzias latipes* in EXTEND 2016. J. Appl. Toxicol. **42**(5), 750–777 (2022). https://doi.org/10.1002/jat.4255
62. C.A. Staples, A. Tilghman Hall, U. Friederich, N. Caspers, G.M. Klecka, Early lifestage and multigeneration toxicity study with bisphenol A and fathead minnows (*Pimephales promelas*). Ecotoxicol. Environ. Saf. **74**(6), 1548–1557 (2011). https://doi.org/10.1016/j.ecoenv.2011.05.010
63. Y. Horie, N. Kanazawa, C. Takahashi, N. Tatarazako, T. Iguchi, Bisphenol A induces a shift in sex differentiation gene expression with testis-ova or sex reversal in Japanese medaka (*Oryzias latipes*). J. Appl. Toxicol. **40**(6), 804–814 (2020). https://doi.org/10.1002/jat.3945
64. F. Zhao, P. Wei, J. Wang, M. Yu, X. Zhang, H. Tian, W. Wang, S. Ru, Estrogenic effects associated with bisphenol a exposure in male zebrafish (Danio rerio) is associated with changes of endogenous 17β-estradiol and gene specific DNA methylation levels. Gen. Comput. Endocrinol. **252**, 27–35 (2017). https://doi.org/10.1016/j.ygcen.2017.07.032
65. B.M. Kim, M.C. Lee, H.M. Kang, J.S. Rhee, J.S. Lee, Genomic organization and transcriptional modulation in response to endocrine disrupting chemicals of three vitellogenin genes in the self-fertilizing fish *Kryptolebias marmoratus*. J. Environ. Sci. (China) **42**, 187–195 (2016). https://doi.org/10.1016/j.jes.2015.08.006
66. W.T. Song, Z.J. Wang, H.C. Liu, Effects of individual and binary mixtures of estrogens on male goldfish (*Carassius auratus*). Fish Physiol. Biochem. **40**(6), 1927–1935 (2014). https://doi.org/10.1007/s10695-014-9980-1
67. L. Sun, X. Lin, R. Jin, T. Peng, Z. Peng, Z. Fu, Toxic effects of bisphenol A on early life stages of Japanese medaka (*Oryzias latipes*). Bull. Environ. Contam. Toxicol. **93**(2), 222–227 (2014). https://doi.org/10.1007/s00128-014-1298-2
68. E. Mihaich, J. Rhodes, J. Wolf, N. van der Hoeven, D. Dietrich, A.T. Hall, N. Caspers, L. Ortego, C. Staples, S. Dimond, S. Hentges, Adult fathead minnow, *Pimephales promelas*, partial life-cycle reproductive and gonadal histopathology study with bisphenol A. Environ. Toxicol. Chem. **31**(11), 2525–2535 (2012). https://doi.org/10.1002/etc.1976
69. V. Hidalgo-Ruz, L. Gutow, R.C. Thompson, M. Thiel, Microplastics in the marine environment: a review of the methods used for identification and quantification. Environ. Sci. Technol. **46**, 3060–3075 (2012). https://doi.org/10.1021/es2031505
70. L. Yang, Y. Zhang, S. Kang, Z. Wang, C. Wu, Microplastics in freshwater sediment: A review on methods, occurrence, and sources. Sci. Total Environ. **754**, 141948 (2021). https://doi.org/10.1016/j.scitotenv.2020.141948
71. Z. Dai, H. Zhang, Q. Zhou, Y. Tian, T. Chen, C. Tu, C. Fu, Y. Luo, Occurrence of microplastics in the water column and sediment in an inland sea

affected by intensive anthropogenic activities. Environ. Pollut. **242**, 1557–1565 (2018). https://doi.org/10.1016/j.envpol.2018.07.131
72. F. Bessa, P. Barría, J.M. Neto, J.P.G.L. Frias, V. Otero, P. Sobral, J.C. Marques, Occurrence of microplastics in commercial fish from a natural estuarine environment. Mar. Pollut. Bull. **128**, 575–584 (2018). https://doi.org/10.1016/j.marpolbul.2018.01.044
73. N. Ohkubo, M. Ito, T. Hano, K. Kono, K. Mochida, Estimation of the uptake and gut retention of microplastics in juvenile marine fish: Mummichogs (Fundulusheteroclitus) and red seabreams (Pagrus major). Mar. Pollut. Bull. **160**, 111630 (2020). https://doi.org/10.1016/j.marpolbul.2020.111630
74. K. Okamoto, M. Nomura, Y. Horie, H. Okamura, Color preferences and gastrointestinal-tract retention times of microplastics by freshwater and marine fishes. Environ. Pollut. **304**, 119253 (2022). https://doi.org/10.1016/j.envpol.2022.119253
75. H. Jacob, M. Besson, P.W. Swarzenski, D. Lecchini, M. Metian, Effects of virgin micro- and nanoplastics on fish: trends, meta-analysis, and perspectives. Environ. Sci. Technol. **54**(8), 4733–4745 (2020). https://doi.org/10.1021/acs.est.9b05995
76. J. He, X. Yang, H. Liu, Enhanced toxicity of triphenyl phosphate to zebrafish in the presence of micro- and nano-plastics. Sci. Total Environ. **756**, 143986 (2021). https://doi.org/10.1016/j.scitotenv.2020.143986
77. J. Wang, X. Li, M. Gao, X. Li, L. Zhao, S. Ru, Polystyrene microplastics increase estrogenic effects of 17α-ethynylestradiol on male marine medaka (Oryzias melastigma). Chemosphere **287**, 132312 (2022). https://doi.org/10.1016/j.chemosphere.2021.132312
78. Y. Takai, H. Tokusumi, M. Sato, D. Inoue, K. Chen, T. Takamura, S. Enoki, Y. Ueno, I.J. Kang, Y. Shimasaki, X. Qiu, Y. Oshima, Combined effect of diazepam and polystyrene microplastics on the social behavior of medaka (Oryzias latipes). Chemosphere **299**, 134403 (2022). https://doi.org/10.1016/j.chemosphere.2022.134403
79. J. Lu, J. Wu, L. Gong, Y. Cheng, Q. Yuan, Y. He, Combined toxicity of polystyrene microplastics and sulfamethoxazole on zebrafish embryos. Environ. Sci. Pollut. Res. Int. **29**(13), 19273–19282 (2022). https://doi.org/10.1007/s11356-021-17198-8
80. ISO/DIS (Draft) Plastics—Marine ecotoxicity testing scheme for soluble decomposition intermediates from biodegradable plastic materials in products intentionally used in the marine environment—Test methods and requirements, ISO/DIS 5430, Draft (2022)
81. Y. Chae, Y.J. An, Effects of micro and nanoplastics on aquatic ecosystems. Current research trends and perspectives. Mar. Pollut. Bull. **124**, 624–632 (2017). https://doi.org/10.1016/j.marpolbul.2017.01.070
82. J.P. da Costa, P.S. Santos, A.C. Duarte, T. Rocha-Santos, Nano plastics in the environment-sources, fates, and effects. Sci. Total Environ. **566**, 15–26 (2016). https://doi.org/10.1016/j.scitotenv.2016.05.041
83. M. Gonzalez-Pleiter, M. Tamayo-Belda, G. Pilido-Reyes, G. Amariei, F. Leganes, R. Rosal, F. Fernandez-Pinas, Secondary nanoplastics released from a biodegradable microplastic severely impact freshwater environments. Environ. Sci: Nano **6**, 1382–1392 (2019). https://doi.org/10.1039/C8EN01427B
84. A. Barrick, O. Champeau, A. Chatel, N. Manier, G. Northcott, L.A. Tremblay, Plastic additives: challenges in ecotox hazard assessment. PeerJ **9**, e11300 (2021). https://doi.org/10.7717/peerj.11300
85. J. Hahladakis, C. Velis, R. Weber, E. Iacovidou, P. Purnell, An overview of chemical additives present in plastics: migration, release, fate and environmental impact during their use, disposal and recycling. J. Hazard Mater. **344**(5), 179–199 (2018). https://doi.org/10.1016/j.jhazmat.2017.10.014
86. OECD, Complementing documents to the emission scenario document on plastic additives: plastic additives during the use of end products. No. 38. ENV/JM/MONO (2019b)

The manufacturer's authorised representative in the EU is Springer Nature Customer Service Centre GmbH, Europaplatz 3, 69115 Heidelberg, Germany. If you have any concerns regarding our products, please contact ProductSafety@springernature.com

Printed and bound by CPI Group (UK) Ltd, Croydon, CR0 4YY
01/10/2026
02225877-0020